QUICK LABS

Teacher's Edition

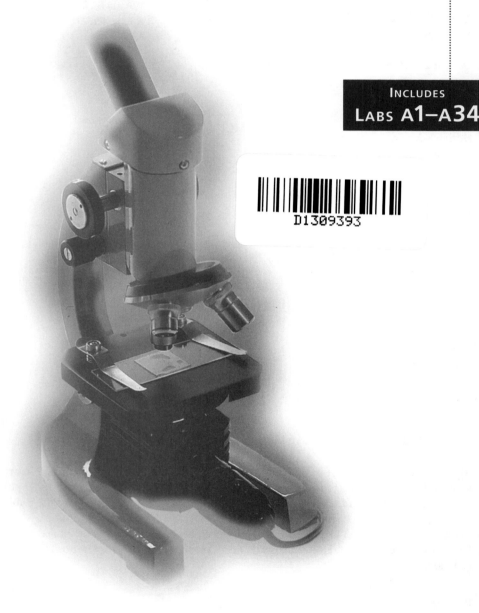

D1309393

HOLT, RINEHART AND WINSTON

Harcourt Brace & Company

Austin • New York • Orlando • Atlanta • San Francisco • Boston • Dallas • Toronto • London

HOLT BIOSOURCES® *LAB PROGRAM*

QUICK LABS

Acknowledgments

Copyediting

Amy Daniewicz
Denise Haney
Steve Oelenberger

Prepress

Rose Degollado

Manufacturing

Mike Roche

Design Development and Page Production

Morgan-Cain & Associates

Cover

Design—Morgan-Cain & Associates
Photography—Sam Dudgeon

Reviewers

Lab Activities
George Nassis
WARD'S Natural Science Establishment
Rochester, NY

Lab Safety
Kenneth Rainis
WARD'S Natural Science Establishment
Rochester, NY

QUICK LABS

Contents

Unit 1 *Cell Structure and Function*

Unit 2 *Genetics*

Unit 3 *Evolution*

Unit 4 *Ecology*

Contents, *continued*

Maintaining a Safe Lab

Building Safety Partnerships: You're Not Alone

A safe laboratory can only be achieved through a partnership among all parties concerned, not just among students or teachers. Materials for teachers and students need to be thorough, explicit, and persuasive. Teachers must actively boost safety consciousness among students, fellow faculty members, administrators, and parents. For success, everyone must agree to respect the same laboratory rules, to obtain and use the proper safety equipment, and to take appropriate precautions during a lab activity.

An excellent way to start building this safety partnership with your students is to use the Safety Contract on the **Holt BioSources Teaching Resources CD-ROM.** Have each student fill out a contract and return it to you. Keep the contracts on file, in case you need to remind students of their promises.

Where to Start

In each lab activity, safety symbols are included and specific safety procedures are highlighted where appropriate. Detailed descriptions of each safety symbol and hazards and precautions related to each one can be found in your biology textbook and in the "Laboratory Safety" section of the pupil's edition of the *Inquiry Skills Development, Laboratory Techniques and Experimental Design,* and *Biotechnology* manuals. The safety symbol descriptions are included in expanded form in this section.

The expanded safety symbol descriptions and other information included in the following sections will help you plan and maintain a safe and healthy laboratory environment.

This information is not all-inclusive. Each school's lab situation is different, and no publication could list safe practices for all situations that could possibly arise.

Be sure that you are aware of any federal, state, or local laws that may cover your lab. Although laws and regulations can vary from place to place and from time to time, you can build a safe program suited to your situation using this information.

Safety Symbols

Eye Safety

- **Wear approved chemical safety goggles as directed.** Goggles should always be worn whenever you or your students are working with a chemical or solution, heating substances, using any mechanical device, or observing physical processes. See the "Safety Equipment" section for specific tips on the types of goggles to be worn.

  Teachers should model appropriate behavior by wearing lab goggles when appropriate. Some teachers have success when they turn this into a game: students found without safety goggles must wear them during the next lecture period. If students catch the teacher not wearing safety goggles, the teacher must wear them during the next lecture period.

- **In case of eye contact, go to an eyewash station and flush eyes (including under the eyelids) with running water for at least 15 minutes.** The teacher or other adult in charge must be notified immediately.

- **Wearing contact lenses for cosmetic reasons is prohibited in the laboratory.** Be sure to make this clear to students at the beginning of the year. First, take a poll of contact-lens wearers. Explain the precautions necessary, noting that liquids or gases can be drawn up under a contact lens and onto the eyeball. If a student must wear contact lenses prescribed by a physician, be sure the students wear approved eyecup safety goggles, which are similar to goggles worn for swimming.

- **Never look directly at the sun through any optical device or lens system, and never gather direct sunlight to illuminate a microscope.** Such actions will concentrate light rays that can severely burn the retina, possibly causing blindness. At the beginning of the year, make sure each microscope you will use has an appropriate and functioning light source.

Electrical Supply

- **Be sure you know the location of the master shut-off for all circuits and other utilities in the lab.** If the circuit breakers are locked up, make sure you have a key in case of an emergency. Color code or label the necessary switches. Be sure you will remember what to do under the pressures of an emergency.

- **Be sure all outlets have correct polarity and ground-fault interruption.** Polarity can be tested with an inexpensive (about $5.00) continuity tester available from most electronic hobby shops. Use only electrical equipment with three-prong plugs and three-wire cords. Each electrical socket in the laboratory must be a three-holed socket with a GFI (ground fault interrupter) circuit. In some cases, rewiring the lab may be necessary. Be sure your supervisors understand the potential hazards and costs of leaving the lab in an unsafe configuration.

- **Electrical equipment should be in the "off" position before it is plugged into a socket.** After a lab activity is completed, the equipment should first be turned off and then unplugged. Wiring hookups should not be made or altered except when an apparatus is disconnected from its AC or DC power source and the power switch, if applicable, is off.

- **Do not let electrical cords dangle from work stations.** Dangling cords are a hazard that can cause tripping or electrical shock.

- **Tape electrical cords to work surfaces.** This will prevent falls and decrease the chances of equipment being pulled off the table. If you find you have too many cords to be taped down, that could be a sign of a poorly designed lab, which will be prone to other problems as well. Sometimes, simply rearranging the lab desks alleviates some flaws.

- **Never use or allow students to use equipment with frayed or kinked cords.** Check all of the electrical equipment at the beginning and end of each year. It is better to omit an activity because the equipment is unsafe than to proceed with an activity that results in an injury to yourself or a student.

- **Never use electrical equipment around water or with wet hands or clothing.** The area under and around electrical equipment should be dry. Electrical cords should not lie in puddles of spilled liquid.

- **Use dry cells or rechargeable batteries as direct current (DC) sources.** Do not use automobile storage batteries or AC to DC converters; these two sources of DC current can present serious electrical shock hazards. When storing dry cells and rechargeable batteries, cover both terminals with insulating tape.

- **Before leaving the laboratory, be sure all electrical equipment has been turned off and is unplugged.**

Clothing Protection

- **Wear a lab apron or lab coat when working in the laboratory to prevent chemicals or chemical solutions from contacting skin or contaminating clothes.** Suggested styles of lab aprons are discussed in the "Safety Equipment" section that follows. Be sure students confine all loose clothing and long jewelry. Open-toed shoes should not be allowed in the laboratory.

Animal Care

- **Do not touch or approach any animal in the wild.** Be sure you and your students are aware of any poisonous or dangerous animals in any area where you will be doing fieldwork.

- **Always insist that students obtain your permission before bringing any animal (or pet) into the school building.** There are legitimate reasons to bring animals to school, but be certain that such an occasion does not present a danger or a distraction to students.

- **Handle all animals with proper caution and respect.** Mishandling or abuse of any animal should not be tolerated. The National Association of Biology Teachers guidelines for the use of live animals, reproduced in the "Animal Care" section, provide a good framework for planning specific procedures.

Sharp Object Safety

- **Use extreme care with all sharp instruments, such as scalpels, sharp probes, and knives.** You may want to consider restricting the use of such objects to lab activities for which there are no substitutes.

- **Never use double-edged razors in the laboratory.**

- **Never cut objects while holding them in your hand.** Place objects on a suitable work surface. Be sure your lab has an adequate supply of dissecting pans and similar surfaces for cutting.

Chemical Safety

More detailed information on chemical hazards, the use of MSDSs (Material Safety Data Sheets), and safe chemical storage can be found in the "Reagents and Storage" and "Chemical Handling and Disposal" sections that follow.

- **Always wear appropriate personal protective equipment.** Eye goggles, gloves, and a lab apron or lab coat should always be worn when working with any chemical or chemical solution.

- **Never taste, touch, or smell any substance or bring it close to your eyes, unless specifically directed to do so.** If students need to note the odor of a substance, have them do so by waving the fumes toward themselves with their hands. Make sure there are enough suction bulbs for any pipetting that needs to be done. Set a good example, and never pipet any substance by mouth.

- **Always handle any chemical or chemical solution with care.** Nothing in the lab should be considered harmless. Even nontoxic substances can easily be contaminated. Check the MSDS for each chemical prior to a lab activity, and observe safe-use procedures. Be sure to have appropriate containers available for unused reagents so that students won't return them to reagent bottles. Store chemicals according to the directions in the "Reagents and Storage" section.

- **Never mix any chemicals unless you are certain about what you are doing and why.** Many common chemicals react violently with each other. Consult section V of a chemical's MSDS for compatibility information.

- **Never pour water into a strong acid or base.** The mixture can produce heat and splatter. Remember this rhyme:

  > "Do as you oughta—
  > Add acid (or base) to water."

- **Have a spill-control plan and kit ready.** Students should not handle chemical spills. Be sure that your spill-control kit contains neutralizing agents, sand, and other absorbent material.

- **Check for the presence of any source of flames, sparks, or heat (open flame, electrical heating coils, etc.) before working with flammable liquids or gases.**

Plant Safety

- **Do not ingest any plant part used in the laboratory (especially seeds sold commercially).** Commercially sold seeds often are coated with fungicidal agents. Do not rub any sap or plant juice on your eyes, skin, or mucous membranes.

- **Wear protective (disposable polyethylene) gloves when handling any wild plant.**

- **Wash hands thoroughly after handling any plant or plant part (particularly seeds).** Avoid touching hands to your face and eyes.

- **Do not inhale or expose yourself to the smoke of any burning plant.** Some irritants travel in smoke and can cause inflammation in the throat and lungs.

- **Do not pick wildflowers or other plants unless permission from appropriate authorities has been obtained in advance.**

Proper Waste Disposal

- **Have students clean and decontaminate all work surfaces and personal protective equipment after each lab activity.** Prompt and frequent cleaning helps keep contamination problems to a minimum.

- **Set aside special containers for the disposal of all sharp objects (broken glass and other contaminated sharp objects) and other contaminated materials (biological or chemical).** Make sure these items are disposed of in an environmentally sound way.

Hygienic Care

- **Keep your hands away from your face and mouth.**

- **Wash your hands thoroughly before leaving the laboratory.** Have bactericidal soap available for students to use.

- **Remove contaminated clothing immediately; launder contaminated clothing separately.** Have a few spare T-shirts and shorts or sweatsuits available in case of an emergency involving clothing.

- **Demonstrate the proper techniques when handling bacteria or microorganisms.** Examine microorganism cultures (such as those in petri dishes) without opening them.

- **Collect all stock and experimental cultures for proper disposal.** See the "Safety With Microbes" section for instructions on materials and cultures used in these lab activities.

Laboratory Safety

Heating Safety

- **When heating chemicals or reagents in a test tube, never point the test tube toward anyone.**

- **Use hot plates, not open flames.** Be sure hot plates have an On-Off switch and indicator light. Never leave hot plates unattended, even for a minute.
  Check all hot plates for malfunctions several times during the school year. Never use alcohol lamps.

- **Know the location of laboratory fire extinguishers and fire blankets.** Have ice readily available in case of burns or scalds. Make certain that your laboratory fire extinguishers are tri-class (A-B-C) and are useful for all types of fires.

- **Use tongs or appropriate insulated holders when heating objects.** Heated objects often do not look hot. Set a good example by using tongs or other holders to handle an object whenever there is a possibility that the object could be warm.

- **Keep combustibles away from heat and other ignition sources.**

Hand Safety

- **Never cut objects while holding them in your hand.**

- **Wear protective gloves when working with stains, chemicals, chemical solutions, or wild (unknown) plants.**

Glassware Safety

- **Inspect glassware before use; never use chipped or cracked glassware.** Use borosilicate glass for heating. Check all glassware several times a year, and discard anything that shows signs of chipping or cracking.

- **Hold glassware firmly, but do not squeeze it.** Glass is fragile and may break if it is not handled carefully. Be sure your hands and the glassware are dry when you are handling glassware.

- **Do not attempt to insert glass tubing into a rubber stopper without taking proper precautions.** Lubricate the stopper and the glass tubing. Use heavy leather gloves to protect your hands from shattering glass. To prevent puncture wounds, be sure your hand is clear of the hole where the glass tubing will emerge.

- **Always clean up broken glass by using tongs and a brush and dustpan.** Discard the pieces in an appropriately labeled "sharps" container.

Safety With Gases

- **Never directly inhale any gas or vapor.** Do not put your nose close to any substance having an odor.

- **Be sure that your lab has excellent ventilation.** Some work will still require a chemical fume hood. If your lab does not have good ventilation, investigate opportunities to improve it. Be certain your supervisors are aware of potential hazards due to ineffective ventilation.

HRW material copyrighted under notice appearing earlier in this work.

Laboratory Rules

Post the following rules in the laboratory, and discuss them with students. Afterward, give students the "Safety Quiz" on the **Holt BioSources Teaching Resources CD-ROM**.

- **Never work alone in the laboratory.**

- **Never perform any experiment not specifically assigned by your teacher.** Never work with any unauthorized material.

- **Never eat, drink, or apply cosmetics in the laboratory.** Never store food in the laboratory. Keep hands away from faces. Wash your hands at the conclusion of each laboratory investigation and before leaving the laboratory. Remember that some hair products are highly flammable, even after application.

- *NEVER* **taste chemicals.** *NEVER* **touch chemicals.** Even common substances should be considered dangerous, since they can be easily contaminated in the lab.

- **Do not wear contact lenses in the lab.** Chemical vapors can get between the lenses and the eyes and cause permanent eye damage.

- **Know the location of all safety and emergency equipment used in the laboratory.** Examples include eyewash stations, safety blankets, safety shower, fire extinguisher, first aid kit, and chemical spill kit.

- **Know fire drill procedures and the locations of exits.**

- **Know the location of the closest telephone,** and be sure there is a posted list of emergency phone numbers, including the poison control center, fire department, police department, and ambulance service.

- **Familiarize yourself with a lab activity—especially safety issues—before entering the lab.** Know the potential hazards of the materials and equipment to be used and the procedures required for the activity. Before you start, ask the teacher to explain any parts you do not understand.

- **Before beginning work: tie back long hair, roll up loose sleeves, and put on any required personal protective equipment as required by your teacher.** Avoid wearing loose clothing or confine loose clothing that could knock things over, ignite from flame, or soak up chemical solutions. Do not wear open-toed shoes to the lab. If there is a spill, your feet could be injured.

- **Report any accidents, incidents, or hazards—no matter how trivial—to your teacher immediately.** Any incident involving bleeding, burns, fainting, chemical exposure, or ingestion should also be reported to the school nurse or physician.

- **In case of fire, alert the teacher and leave the laboratory.**

- **Keep your work area neat and uncluttered.** Bring only the books and materials needed to conduct a lab activity. Stay at your work area as much as possible. The less movement in a lab, the fewer spills and other accidents that can occur.

- **Clean your work area at the conclusion of a lab activity as your teacher directs.**

- **Wash your hands with soap and water after each lab activity.**

Safety Equipment

Do You Have What It Takes?

- **Chemical goggles** (meeting ANSI [American National Standards Institute] standard Z87.1): These should be worn when working with any chemical or chemical solution other than water, when heating substances, when using any mechanical device, or when observing physical processes that could eject an object.

  Wearing contact lenses for cosmetic reasons should be prohibited in the laboratory. If a student must wear contact lenses prescribed by a physician, that student should wear eyecup safety goggles meeting ANSI standard Z87.1 (similar to swimmers' goggles).

- **Face shield** (meeting ANSI standard Z87.1): Use in combination with eye goggles when working with corrosives.

- **Eyewash station:** The station must be capable of delivering a copious, gentle flow of water to both eyes for at least 15 minutes. **Portable liquid supply devices are not satisfactory and should not be used.** A plumbed-in fixture or a perforated spray head on the end of a hose attached to a plumbed-in outlet and designed for use as an eyewash fountain is suitable if it meets ANSI standard Z358.1 and is within a 30-second walking distance from any spot in the room.

- **Safety Shower** (meeting ANSI standard Z358.1): Location should be within a 30-second walking distance from any spot in the room. Students should be instructed in the use of the safety shower for a fire or chemical splash on their body that cannot be simply washed off.

- **Gloves:** Polyethylene, neoprene, or disposable plastic may be used. Nitrile or butyl rubber gloves are recommended when handling corrosives.

- **Apron:** Gray or black rubber-coated cloth or a nylon-coated vinyl halter is recommended.

Prudent Precautions

What would you do if a student dropped a liter bottle of concentrated sulfuric acid? RIGHT NOW? Are you prepared? Could you have altered your handling and storage methods to prevent or lessen the severity of this incident? PLAN now how to effectively react BEFORE you need to. Planning tips include the following:

1. Post the phone numbers of your regional poison control center, fire department, police department, ambulance service, and hospital ON your telephone.

2. Practice fire and evacuation drills during labs and at all times during the year, not just in the fall. Post an evacuation diagram and an established evacuation procedure by every entrance to the laboratory.

3. Have drills on what students MUST do if they are on fire or experience chemical contact or exposure.

4. Mark the locations of eyewash stations, the safety shower, fire extinguishers (A-B-C tri-class), the chemical spill kit, the first aid kit, and fire blankets in the laboratory and storeroom. Make sure you have all of the necessary safety equipment prior to conducting each lab activity, and be certain the equipment is in good working order.

5. Lock your laboratory (and storeroom) when you are not present.

6. Compile an MSDS file for all chemicals. This reference resource should be readily accessible in case of spills or other incidents. (Information about MSDSs is found in the "Reagents and Storage" section.)

7. Develop spill-control procedures. Handle only incidents that you FEEL COMFORTABLE handling. Situations of greater severity should be handled by trained hazardous-material responders.

8. Under no circumstances should students fight fires or handle chemical spills.

9. Be sure to recognize and heed the signal words used on most safety labels for materials, equipment, and procedures:
CAUTION—low level of risk associated with use or misuse
WARNING—moderate level of risk associated with use or misuse
DANGER—high level of risk associated with use or misuse

10. Be trained in first aid and basic life support (CPR) procedures. Have first aid kits and spill kits readily available.

11. Before the class begins a lab activity, review specific safety rules and demonstrate proper procedures.

12. Never permit students to work in your laboratory without your supervision. No unauthorized investigations should ever be conducted, nor should unauthorized materials be brought into the laboratory.

13. Fully document ANY INCIDENT that occurs. Documentation will provide the best defense in terms of liability, and it is a critical tool in helping to identify area(s) of laboratory safety that need improvement. Remind students that any safety incident, no matter how trivial, must be reported directly to you.

Safety With Microbes

What You Can't See CAN Hurt You

Pathogenic (disease-causing) microorganisms are not appropriate investigation tools in the high school laboratory and should never be used.

Consult with the school nurse to screen students whose immune system may be compromised by illness or may be receiving immunosuppressive drug therapy. Such individuals are extraordinarily sensitive to potential infection from generally harmless microorganisms and should not participate in laboratory activities unless permitted to do so by a physician. Do not allow students with any open cuts, abrasions, or open sores to work with microorganisms.

Aseptic Technique

Demonstrate correct aseptic technique to students PRIOR to conducting a lab activity. Never pipet liquid media by mouth. Wherever possible, use sterile cotton applicator sticks in place of inoculating loops and Bunsen burner flames for culture inoculation. Remember to use appropriate precautions when disposing of cotton applicator sticks: they should be autoclaved or sterilized before disposal.

Treat ALL microbes as pathogenic. Seal with tape all petri dishes containing bacterial cultures. Do not use blood agar plates, and never attempt to cultivate microbes from a human or animal source.

Never dispose of microbe cultures without first sterilizing them. Autoclave or steam-sterilize all used cultures and any materials that have come in contact with them at 120°C and 15 psi for 15–20 minutes. If an autoclave or steam sterilizer is not available, flood or immerse these articles with full-strength household bleach for 30 minutes, and then discard. Use the autoclave or steam sterilizer yourself; do not allow students to use these devices.

Wash all lab surfaces with a disinfectant solution before and after handling bacterial cultures.

Handling Bacteriological Spills

Never allow students to clean up bacteriological spills. Keep on hand a spill kit containing 500 mL of full-strength household bleach, biohazard bags (autoclavable), forceps, and paper towels.

In the event of a bacterial spill, cover the area with a layer of paper towels. Wet the paper towels with the bleach, and allow to stand for 15–20 minutes. Wearing gloves and using forceps, place the residue in the biohazard bag. If broken glass is present, use a brush and dustpan to collect the broken material and place it in a suitably marked container.

Preserved Specimens

Fixatives Do not Distinguish Between Living and Dead Tissue

The following practices are recommended when handling or dissecting any preserved specimen:

1. NEVER dissect road kills or nonpreserved slaughterhouse material. Doing so increases the risk of infection.

2. Wear protective gloves and splashproof safety goggles at all times when handling preserving fluids and preserved specimens and during dissection.

3. Wear lab aprons. Use of an old shirt or smock is recommended.

4. Conduct dissection activities in a well-ventilated area.

5. Do not allow preservation or body cavity fluids to come into contact with skin. Fixatives do not distinguish between living and dead tissue.

Biological supply firms use formalin-based fixatives of varying concentration to initially fix zoological and botanical specimens. Generally, it is the practice to post-treat and ship specimens in holding fluids or preservatives that do not contain formalin.

Ward's Natural Science Establishment provides specimens that are freeze-dried and rehydrated in a 10 percent isopropyl alcohol solution. In these specimens, no other hazardous chemical is present.

Many suppliers provide fixed botanical materials in 50 percent glycerin.

Reduction of Free Formaldehyde

Current federal regulations mandate a permissible formaldehyde exposure level of 0.75 ppm. Contact your supplier for an MSDS that details the amount of formaldehyde present as well as any gas-emitting characteristics for individual specimens.

Prewash specimens (in a loosely covered container) in running tap water for 1 to 4 hours to dilute the fixative. Formaldehyde may also be chemically bound by immersing washed specimens in a 0.5–1.0 percent potassium bisulfate solution overnight or by placing them in holding solutions containing 1 percent phenoxyethanol.

Animal Care

Ethics in the Laboratory

It is recommended that teachers follow the *Guidelines for the Use of Live Animals* established by the National Association of Biology Teachers. The "Chemical Handling and Disposal" section provides specific guidelines for taking care of animals after lab activities.

NABT GUIDELINES FOR THE USE OF LIVE ANIMALS

(Revised April 1991)

Living things are the subject of biology, and their direct study is an appropriate and necessary part of biology teaching. Textbook instruction alone cannot provide students with a basic understanding of life and life processes. The National Association of Biology Teachers recognizes the importance of research in understanding life processes and providing information on health, disease, medical care, and agriculture.

The abuse of any living organism for experimentation or any other purpose is intolerable in any segment of society. Because biology deals specifically with living things, professional biology educators must be especially cognizant of their responsibility to prevent the inhumane treatment of living organisms in the name of science and research. This responsibility should extend beyond the confines of the teacher's classroom to the rest of the school and community.

The National Association of Biology Teachers believes that students learn the value of living things, and the values of science, by the events they witness in the classroom. Such teaching activities should develop in students and teachers a sense of respect and pleasure in studying the wonders of living things. NABT is committed to providing sound biological education and promoting humane attitudes toward animals. These guidelines should be followed when live animals are used in the classroom:

A. Biological experimentation should be consistent with a respect for life and all living things. Humane treatment and care of animals should be an integral part of any lesson that includes living animals.

B. Exercises and experiments with living things should be within the capabilities of the students involved. The biology teacher should be guided by the following conditions:

  1. The lab activity should not cause the undue loss of a vertebrate's life. Bacteria, fungi, protozoans, and invertebrates should be used in activities that may require use of harmful substances or loss of an organism's life. These activities should be clearly supported by an educational rationale and should not be used when alternatives are available.

2. A student's refusal to participate in an activity (e.g., dissection or experiments involving live animals, particularly vertebrates) should be recognized and accommodated with alternative methods of learning. The teacher should work with the student to develop an alternative for obtaining the required knowledge or experience. The alternative activity should require the student to invest a comparable amount of time and effort.

C. Vertebrate animals can be used as experimental organisms in the following situations:

1. Observations of normal living patterns of wild animals in their natural habitat or in zoological parks, gardens, or aquaria.

2. Observations of normal living functions such as feeding, growth, reproduction, activity cycles, etc.

3. Observations of biological phenomenon among and between species such as communication, reproductive and life strategies behavior, interrelationships of organisms, etc.

D. If live vertebrates are to be kept in the classroom, the teacher should be aware of the following responsibilities:

1. The school, under the biology teacher's leadership, should develop a plan on the procurement and ultimate disposition of animals. Animals should not be captured from or released into the wild without the approval of both a responsible wildlife expert and a public health official. Domestic animals and "classroom pets" should be purchased from licensed animal suppliers. They should be healthy and free of diseases that can be transmitted to humans or to other animals.

2. Animals should be provided with sufficient space for normal behavior and postural requirements. Their environment should be free from undue stress such as noise, overcrowding, and disturbance caused by students.

3. Appropriate care—including nutritious food, fresh water, clean housing, and adequate temperature and lighting for the species—should be provided daily, including weekends, holidays, and long school vacations.

4. Teachers should be aware of any student allergies to animals.

5. Students and teachers should immediately report to the school health nurse all scratches, bites, and other injuries, including allergies or illnesses.

6. There should always be supervised care by a teacher competent in caring for animals.

E. Animal studies should always be carried out under the direct supervision of a biology teacher competent in animal care procedures. It is the responsibility of the teacher to ensure that the student has the necessary comprehension for the study. Students and teachers should comply with the following:

1. **Students should not be allowed to perform surgery on living vertebrate animals.** Hence, procedures requiring the administration of anesthesia and euthanasia should not be done in the classroom.

2. Experimental procedures on vertebrates should not use pathogenic microorganisms, ionizing radiation, carcinogens, drugs or chemicals at toxic levels, drugs known to produce adverse or teratogenic effects, pain-causing drugs, alcohol in any form, electric shock, exercise until exhaustion, or other distressing stimuli. No experimental procedures should be attempted that would subject vertebrate animals to pain or distinct discomfort, or interfere with their health in any way.

3. Behavioral studies should use only positive-reinforcement techniques.

4. Egg embryos subjected to experimental manipulation should be destroyed 72 hours before normal hatching time.

5. Exceptional original research in the biological or medical sciences involving live vertebrate animals should be carried out under the direct supervision of an animal scientist, e.g., an animal physiologist, or a veterinary or medical researcher, in an appropriate research facility. The research plan should be developed and approved by the animal scientist and reviewed by a humane society professional staff person prior to the start of the research. All professional standards of conduct should be applied, as well as humane care and treatment, and concern for the safety of the animals involved in the project.

6. Students should not be allowed to take animals home to carry out experimental studies.

F. Science fair projects and displays should comply with the following:

1. The use of live animals in science fair projects shall be in accordance with the above guidelines. In addition, no live vertebrate animals shall be used in displays for science fair exhibitions.

2. No animal or animal products from recognized endangered species should be kept and displayed.

From "NABT Guidelines for the Use of Live Animals." Copyright © 1991 by National Association of Biology Teachers. Reprinted by permission of *National Association of Biology Teachers, Reston, VA 22090.*

Reagents and Storage

General Guidelines

- Store bulk quantities of chemicals in a safe and secure storeroom, not in the teaching laboratory. Store them in well-ventilated, dry areas protected from sunlight and localized heat. Store by similar hazard characteristics, not alphabetically. (See "Chemical Hazard Classes" and "Chemical Storage" for additional recommendations.)

- Label student reagent containers with the substance's name and hazard class(es). Be sure to use labeling materials that won't be affected by the reagent or other chemicals that will be stored nearby. (See "Chemical Hazard Classes" for additional recommendations.)

- Dispose of hazardous waste chemicals according to federal, state, and local regulations. Refer to the Material Safety Data Sheets (available through your supplier) for recommended disposal procedures. Some disposal information is also included in the "Chemical Handling and Disposal" section on page T25. **NEVER ASSUME** that a reagent can be safely poured down the drain.

- Have a chemical-spill kit immediately available. Know the procedures for handling a spill of any chemical used during a lab activity or in preparing reagents. Never allow students to clean up hazardous chemical spills.

- Remove all sources of flames, sparks, and heat from the laboratory when any flammable material is being used.

Chemical Record-Keeping and MSDSs

Maintaining a current inventory of chemicals can help you keep track of purchase dates and amounts. In addition, if you use your inventory to determine what you need for the school year, you can cut down on storage problems.

The purpose of a Material Safety Data Sheet (MSDS) is to provide readily accessible information on chemical substances commonly used in the science laboratory or in industry. MSDSs are available from suppliers of chemicals.

MSDSs should be kept on file and referred to BEFORE handling ANY chemical. The MSDSs can also be used to instruct students on chemical hazards, to evaluate spill and disposal procedures, and to warn of incompatibilities with other chemicals or mixtures.

Each MSDS is divided into the following sections:

I. Material Identification: includes name, common synonyms, reference codes, and precautionary labeling

II. Ingredients and Hazards: identifies dangerous components of mixtures

III. Physical Data: includes information such as melting point, boiling point, appearance, odor, density, etc.

IV. Fire and Explosion Hazard Data: includes flash point, description of fire-extinguishing media and procedures, and information on unusual fire and explosion hazards

V. Health Hazard Data: describes problems associated with inhalation, skin contact, eye contact, skin absorption, and ingestion, along with first aid procedures

VI. Reactivity Data: includes information on incompatible types of chemicals and likely decomposition products

VII. Spill, Leak, and Disposal Procedures: includes step-by-step information

VIII. Special Protection Information: describes equipment needed for safe use

IX. Special Precautions and Comments: describes storage requirements and other notes

WARD'S has a pocket guide (Ward's Catalog No. 32 M 0002) that explains in greater detail how to use Material Safety Data Sheets.

Chemical Hazard Classes

The hazards presented by any chemical can be grouped into the following categories. (It is important to keep in mind that a particular chemical may have more than one of these hazards.)

- **Flammable**
- **Corrosive**
- **Poisonous (toxic)**
- **Reactive**

A fifth category, those chemicals that do not possess the above properties, are termed "low hazard" materials. Water is an example of a low hazard material. Although these materials may not ordinarily represent a hazard, their presence in the lab requires that they be treated differently than they would be in a kitchen or backyard.

The following precautions should be used with all hazard classes:

- Require the use of safety goggles, gloves, and lab aprons.
- Minimize the amounts available in the lab (100 mL or less).
- Become familiar with first aid measures for each chemical used.
- Become familiar with incompatibility issues for each chemical used.
- Emphasize how essential lab and storeroom cleanliness and personal hygiene are when dealing with hazardous materials.
- Keep hazardous chemicals in approved containers that are kept closed and stored away from sunlight and rapid temperature changes.
- Store chemicals of each hazard class away from those in other hazard classes.
- Keep a designated and locked storage cabinet for each hazard class.

FLAMMABLE

Prevention and Control Measures

- Store away from oxidizers and reactives.
- Keep containers closed when not in use.

- The ignition source is the easiest of the three components to remove. Check for presence of lighted burners, sources of sparks (including static charge, friction, and electrical equipment), and hot objects such as hot plates or incandescent bulbs.
- Ground (electrically) all bulk metal containers when dispensing flammable liquids.
- Flammable vapors are usually heavier than air and can travel considerable distances before being diluted below ignitable concentrations.
- Ensure that there are class B fire extinguishers present in the laboratory and store room.
- Students should be drilled in EXACTLY what they must do if their clothes or hair catches fire. Practice "drop and roll" techniques. Both a safety shower and fire blankets should be available. Inform students that the shower is the best way to put out a fire on polyester clothing.
- Conduct a fire inspection with members of the local fire department at least once a year. Practice fire drills regularly.
- Provide adequate ventilation.
- Prepare for spills by having absorbent, vapor-reducing materials (available commercially) close at hand. Plan to have enough absorbent material to handle the maximum volume of flammable substances on hand.

Additional Protective Equipment

- Nitrile or butyl rubber gloves
- Approved storage containers
- Face shield (recommended)
- Fire blanket
- Safety shower
- Fire extinguishers (Class B)

CORROSIVE

Prevention and Control Measures

- Always wear a face shield along with eye goggles when handling solutions of any corrosive material with concentrations greater than 1 mol/liter.
- Have an eyewash station in close proximity.
- Wear the correct type of hand protection that will be impervious to the corrosive being handled. (Nitrile or butyl rubber gloves are generally recommended.)
- Provide adequate ventilation.
- Prepare for spills by having neutralizing reagents close at hand and in sufficient quantities for materials on hand.

Additional Protective Equipment

- Nitrile or butyl rubber gloves
- Safety shower
- Face shield
- Eyewash stations
- Sleeve gauntlets

Laboratory Safety

POISONOUS (TOXIC)
Prevention and Control Measures

- Treat all chemicals as toxic until proven otherwise. Above all, emphasize barriers, cleanliness, and avoidance of contact when handling any chemical.
- Wear protective equipment over exposed skin areas and eyes.
- Handle all contaminated glass and metal carefully. Remember that any sharp object can be a vehicle for introducing a toxic substance.
- Provide good ventilation. Use a chemical fume hood if possible.
- Recognize symptoms of overexposure and typical routes of introduction for each chemical used during a lab activity.
- Remember that the skin is not a good barrier to many toxic chemicals.
- Post the phone number of the nearest poison control center ON your phone.

Additional Protective Equipment

- Container for sharp objects
- Chemical fume hood
- Particle face mask

Chemical Storage

Never store chemicals alphabetically, as that greatly increases the risk of a violent reaction. Take these additional precautions.

1. Always lock the storeroom and all cabinets when not in use.
2. Do not allow students in the storeroom or preparation areas.
3. Avoid storing chemicals on the floor of the storeroom.
4. Do not store chemicals above eye level or on the top shelf in the storeroom.
5. Be sure shelf assemblies are firmly secured to walls.
6. Provide antiroll lips for all shelves.
7. Use shelving constructed of wood. Metal cabinets and shelves are easily corroded.
8. Avoid metal adjustable shelf supports and clips. They can corrode, causing shelves to collapse.
9. Store acids in their own locking storage cabinet.
10. Store flammables in their own locking storage cabinet.
11. Store poisons in their own locking storage cabinet.
12. Store oxidizers by classification, preferably in their own locking storage cabinets.

Additional Resources

Your school district may have more information on safety issues. Some districts have a safety officer responsible for safety throughout their schools. Other possible sources for information include your state education agency, teachers' and science teachers' associations, and local colleges or universities.

The **American Chemical Society Health and Safety Service** will refer inquiries about health and safety to appropriate resources.

American Chemical Society (ACS)
1155 Sixteenth Street, N.W.
Washington, D.C. 20036
(202) 872-4511

Hazardous Materials Information Exchange (HMIX)

Sponsored by the Federal Emergency Management Agency and the United States Department of Transportation, HMIX serves as a reliable on-line database. It can be accessed through an electronic bulletin board, and it provides information regarding instructional material and literature listings, hazardous materials, emergency procedures, and applicable laws and regulations.

HMIX can be accessed by a personal computer with a modem. Dial (312) 972-3275. The bulletin board is available 24 hours a day, seven days a week. The service is available free of charge. You pay only for the telephone call.

Safety Reference Works

Gessner, G. H., ed. *Hawley's Condensed Chemical Dictionary* (11th ed.). New York: Van Nostrand Reinhold, 1987.

A Guide to Information Sources Related to the Safety and Management of Laboratory Wastes from Secondary Schools. New York State Environmental Facilities Corp., 1985.

Lefevre, M. J. *The First Aid Manual for Chemical Accidents.* Stroudsberg, PA: Dowdwen, 1989.

Pipitone, D., ed. *Safe Storage of Laboratory Chemicals.* New York: John Wiley, 1984.

Prudent Practices for Disposal of Chemicals from Laboratories. Washington, D.C.: National Academy Press, 1983.

Prudent Practices for Handling Hazardous Chemicals in Laboratories. Washington, D.C.: National Academy Press, 1981.

Strauss, H., and M. Kaufman, eds. *Handbook for Chemical Technicians.* New York: McGraw-Hill, 1981.

WARD'S MSDS Database and User's Guide. WARD'S CD-ROM, Catalog Number 74M5070.

Release and Disposal of Organisms

The ultimate responsibility falls on the teacher in assuring that each organism brought into the classroom receives adequate care during its stay, release, and final disposition.

The following information is provided regarding the organisms used in the *Quick Labs* to aid you in making informed decisions regarding these matters.

General Guidelines

- Insects—Permits are required to possess or release certain insects (cockroaches and termites). Check with your local office of the Animal and Plant Health Inspection Service (APHIS), U.S. Department of Agriculture, or contact WARD'S.

    Crickets—Release not recommended.

- Microbes—Bacteria, fungi, yeast, and growth media or materials that have come into contact with these organisms should not be discarded without prior decontamination (sterilization). See: *Safety With Microbes in Guidelines For Laboratory Safety.*

    Physarum   Yeast

- Microinvertebrates/Protists—These may be freely released in aquatic environments. Do not release nematodes unless specifically directed to do so.

    Euglena   Paramecium   Pondwater

- Macroinvertebrates/Vertebrates—Exotic or nonindigenous forms should not be released into native habitats. In many cases, these organisms may not survive climatic conditions or may interfere with native fauna and flora. Contact your local APHIS office or WARD'S for specific information about whether a particular organism would be considered nonindigenous to *your* area. In some cases, you may be able to find a home for an animal at a local pet shop.

    Chameleon (Anole)—release only in South and Southeast U.S.
    Earthworms— release to appropriate terrestrial habitat
    Goldfish—exotic; do not release
    Grassfrog (adult/tadpole/egg)—release only to appropriate aquatic habitat

- Plants—Locally cultivated native plants may be introduced by replanting. Ornamental plants that are not cultivated locally should not be introduced into native habitat. Some states (California) have strict rules regarding the procurement or introduction of non-indigenous plants because of the possible presence of root-damaging nematodes. Usually, plants shipped to these states must pass inspection.

    Ceratoperis —release to aquatic environment not recommended
    Coleus —may be replanted if available locally
    Ferns—replanting not recommended
    Polytricium (moss) —replanting not recommended
    Tradescantia (Wandering Jew)—replanting not recommended
    Venus' Flytrap—replanting not recommended

Chemical Handling and Disposal

This information is furnished without warranty of any kind. Teachers should use it only as a supplement to other information they have and make independent determinations of its suitability and completeness as it relates to their own district guidelines.

NONHAZARDOUS CHEMICALS

The following materials are considered nonhazardous; they do not meet the published criteria of established hazard characteristics or are not specifically regulated as hazardous substances.

- **STORAGE:** GREEN (General Storage)
- **DISPOSAL:** LOW HAZARD for laboratory handling. Avoid creating dusts when working with these materials. May be disposed as an inert solid or liquid waste.

Detain™ (protist slowing agent)	Sand, fine
Food coloring	Sugar
Pumice	Vegetable oil

HAZARDOUS CHEMICALS

Before using or disposing of any of the materials listed below, familiarize yourself with the safety and handling procedures and storage information listed in *Guidelines For Laboratory Safety*. Also refer to individual reagent labels and material Safety Data Sheets for further information about hazards and precautions.

Container Labeling

Assure that each container used by students in the laboratory is properly labeled with the following information:

- the name of the material and its concentration (if a solution)
- the names of individual components and their respective concentrations (if a mixture)
- the appropriate SIGNAL WORD
- a declarative statement of potential hazard or hazards
- immediate first aid measures

> Example:
>
> **Lugol's Iodine Solution**
> **WARNING: Poison if Ingested/Irritant**
> Do not ingest. Avoid skin/eye contact.
> Flush spills and splashes with water for 15 minutes; rinse mouth with water.
> Call your teacher immediately.

You should be aware of local, state, and federal regulations governing the disposal of hazardous materials. Contact a licensed Treatment—Storage—Disposal (TSD) facility for disposal of bulk or large quantities of hazardous chemicals. Disposal protocols outlined below are ONLY for the substances (and quantities) specified.

Unless your school's drains are connected to a sanitary sewer system, no chemicals should ever be introduced to the drain. Never pour chemicals or reagents down the drain if you have a septic system. Even if you are connected to a sanitary sewer, do not dispose of any chemical down the drain unless you are certain it is safe and permitted.

The chemicals and reagents are classified by a color code to indicate hazard level. Remember to store chemicals of different hazard classes away from each other. YELLOW: Reactives, RED: Flammables, WHITE: Corrosives, BLUE: Toxics (poisons), GREEN: Low Hazard for laboratory use.

Preserved Materials—WARDsafe holding solution

- **CAUTION:** Aqueous solution may contain traces of original fixative. Prolonged contact may be irritating to skin and eyes and may cause allergic reaction in hypersensitive individuals. Discontinue use if redness or swelling occurs. In case of contact, flush with water, including under eyelids, for 15 minutes. Contact physician if irritation or redness persists. May be toxic if swallowed—contains 2.8% METHYL ALCOHOL. If conscious, drink 8–10 oz (240–300 mL) water to dilute material; INDUCE VOMITING. Get prompt medical attention.
- **PPE:** Chemical safety goggles; polyethylene gloves; apron/smock
- **STORAGE:** GREEN (general storage)
- **DISPOSAL:** Wear PPE. Dilute small volumes (up to 1 gal) 1:20 with water in bucket. Place bucket in sink and run water to overflowing for 10 minutes, flushing into a sanitary sewer. Specimens in WARDsafe may be disposed of as inert solid waste if they are damp (not wet).

Preserved Materials—Grasshoppers

- **CAUTION:** Alcoholic solution of WARDsafe containing 20% isopropyl alcohol. Prolonged contact may be irritating to skin and eyes; may cause allergic reaction in hypersensitive individuals. Discontinue use if redness or swelling occurs. In case of contact: flush with water, including under eyelids, for 15 minutes. Contact physician if irritation or redness persists. May be toxic if swallowed. If conscious, drink 8–10 oz (240-300 mL) water to dilute material; INDUCE VOMITING. Get prompt medical attention.
- **PPE:** Chemical safety goggles; polyethylene gloves; apron/smock
- **STORAGE:** GREEN (general storage)
- **DISPOSAL:** Wear PPE. Dilute small volumes (up to 500 mL) 1:20 with water in a beaker. Place beaker in sink and run water to overflowing for 10 minutes, flushing into a sanitary sewer. Specimens in WARDsafe may be disposed of as inert solid waste if they are damp (not wet).

Seeds

Broad Bean, Broad Windsor Bean, Oriental Mung Bean

- **CAUTION:** Seeds are usually treated with the fungicide CAPTAN [80% *N*-(trichloromethyl)-4-cyclohexene-1,2-dicarboximide]. POISON—do not ingest; avoid skin contact. Do not burn.
- **PPE:** Polyethylene gloves
- **STORAGE:** GREEN (general storage)
- **DISPOSAL:** Wear PPE. Place in sealed container so that seeds cannot be ingested by animals.

Starch Agar Plates

- **PREPARATION:** Mix 25 g starch agar in 1 L distilled water. Bring mixture to a boil while stirring. Autoclave at 120°C at 15–18 psi. Dispense 10–15 mL per plate using sterile technique.
- **STORAGE:** GREEN (general storage). Store under refrigeration (6°C) until needed.
- **DISPOSAL:** LOW HAZARD for laboratory handling without bacterial contamination.

Master Materials List, by Category

This materials cross-reference guide was prepared by WARD'S Natural Science, the exclusive science supplier for Holt, Rinehart and Winston, publisher of the Holt BioSources Lab Program.

Materials are grouped into four categories: Biological Supplies, Chemicals and Media, Laboratory Equipment, and Miscellaneous. Each entry is listed alphabetically, followed by the WARD'S catalog number and package size. The second column gives the lab number in which the material is used. A designation of *local* means that the item should be available locally.

WARD'S also has available a convenient and easy computer software ordering system specifically designed for use with the Holt BioSources Lab Program. The software ordering system lists all required and supplemental materials per group or class size needed for every lab. "Click" on the products you need and the software automatically creates your "shopping list," keeping track of the materials you ordered and their costs. The software ordering system is available for both Macintosh and IBM computers.

To order, or for questions concerning the use of WARD'S materials, call toll-free 1-800-962-2660 or fax WARD'S at 1-800-635-8439.

WARD'S Natural Science Establishment, Inc.
5100 W. Henrietta Road
P.O. Box 92912
Rochester, NY 14692-9012

Biological Supplies

Item	Lab
Animal phyla collection, preserved, set of 10 (62 T 0065)	**A23**
Animal survey set, preserved, each/10 (62 T 0010)	**A21**
Begonia, living, each (86 T 7015)	**A18**
Broad bean seed, .5 lb pkg (86 T 8003)	**A20**
Broad windsor bean seed, pkt of 30 (86 T 8000)	**A10**
Ceratopteris, living, each (86 T 7600)	**A18**
Chameleons, American living, pkg of 3 (87 T 8135)	**A10**
Coleus plant, living, each (86 T 6800)	**A2, A18**
Crickets, living, pkg of 10 (87 T 6100)	**A24**
Crustose lichens (SECT) QS Slide, each (91 T 3980)	**A2**
Earthworms (*Lumbricus*), living, pkg of 10 (87 T 4660)	**A2**
Elrathia kingi, large fossil, each (53 T 2920)	**A2**
Euglena sp culture, cul (87 T 0100)	**A17**

Item	Lab
Fantail goldfish, living, pkg of 6 (87 T 8116)	**A2, A10**
Fern spores, hardy mixed, pkt (86 T 5501)	**A10**
Fern, woodland, living, each (86 T 5500)	**A10**
Frog dev. state, 1-cell, pres., vl/10 (69 T 2208)	**A10, A22**
Frog eggs (limited availability), pkg of 100 (87 T 8205)	**A22, A34**
Grassfrog skeleton, each (65 T 2210)	**A26**
Grassfrogs, small living, pkg of 12 (87 T 8217)	**A25**
Grasshopper, *Lubber,* preserved, 10/jar (68 T 4052)	**A2**
Human blood (SM) GS Slide, each (93 T 6541)	**A32**
Moss (*Polythichum*), living (86 T 4360)	**A10, A18**
Oriental mung bean seed, pkt of 100 (86 T 8007)	**A2**
Paramecium caudatum culture (87 T 1310)	**A17**

Item	Lab
Perch skeleton, each (65 T 1550)	**A26**
Physarum polychephalum culture (85 T 4751)	**A2**
Pigeon skeleton, each (65 T 4470)	**A26**
Pondwater, 1 gal (88 T 7010)	**A34**
Protist slowing agent: Detain, .5 oz (37 T 7950)	**A17**
Skeleton, plastic rod mt., each (82 T 3016)	**A26**
Skull, plastic human, each (82 T 3031)	**A27**
Snake skeleton, each (65 T 3500)	**A26**
Starch agar plates, pkg of 6 (88 T 0927)	**A2**
Sweet corn seed, untreated, 2 oz pkt (86 T 8080)	**A20**
Syringa (lilac) leaf (CS) QS slide, each (91 T 8272)	**A19**
Tradescantia, living, each (86 T 7300)	**A18**
Turtle skeleton, each (65 T 3330)	**A26**
Venus' flytrap, living, each (86 T 7401)	**A18**
Yeast, viable, 10 g pkt (88 T 0929)	**A2, A17**

Chemicals and Media

Item	Lab
Aluminum foil, 12 in. wide roll, 25 ft roll (15 T 1009)	**A16, A24**
Food coloring, 4 colors, pkg of 4 (15 T 0071)	**A4**
Pumice–Italy, SS, pkg of 10 (47 T 6447)	**A2**
Safranin O 1% solution, 120 mL (38 W 7018)	**A17**
Sand, fine white, bag of 500 g (20 T 7423)	**A3**

Laboratory Equipment

Item	Lab
Apron, disposable polyethylene, box of 100 (15 T 1050)	**A4, A14, A16, A21, A23, A34**
Beaker, low-form 1000 mL Griffin, each (17 T 4080)	**A16, A25**
Beaker, low-form 100 mL Griffin, each (17 T 4020)	**A34**
Beaker, low-form 250 mL Griffin, each (17 T 4040)	**A17, A25**
Beaker, low-form 400 mL Griffin, each (17 T 4050)	**A1, A4**
Beaker, low-form 600 mL Griffin, each (17 T 4060)	**A16, A24**
Corks, size 2, pkg of 100 (15 T 8362)	**A2**
Coverslips, 22 mm plastic, box of 100 (14 T 3555)	**A2, A17**
Dish, specimen 4 1/2 × 2 3/4 in., each (17 T 0550)	**A22, A34**

Item	Lab
Dish, specimen 8 1/4 × 3 1/4 in., each (17 T 0560)	**A13**
Dissection pan set (18 T 3665)	**A21, A23**
Forceps, dissecting, medium, each (14 T 1001)	**A22**
Funnel, 58 mm polypropylene, bag of 100 (18 T 1300)	**A3**
Gloves, disposable, medium, box of 100 (15 T 1071)	**A14, A16, A21, A23, A34**
Gloves, heat defier kelnit cotton, pr (15 T 1095)	**A4**
Graduated cylinder, glass 100 mL, each (17 T 0173)	**A3**
Hot plate, (700 W) single burner, each (15 T 7999)	**A4, A16, A17, A34**
Incubator, lab, each (15 T 0060)	**A16**
Lamp, with reflector and clamp, each (36 T 4168)	**A24**
Lung volume bag set, (14 T 5051)	**A31**
Magnifier, dual 3×, 6×, each (24 T 1112)	**A22**
Meterstick (15 T 4065)	**A31**
Microscope slide, precleaned, pkg of 72 (14 T 3500)	**A2, A17**
Microscope, each (24 T 2310)	**A2, A17, A19, A32**
Pen, black wax marker for glass, each (15 T 1155)	**A1, A4, A34**
pH paper, 1-14 range, vl/100 (15 T 2558)	**A16**
Pins, T nickel plated, pkg of 100 (14 T 0201)	**A28**
Pipet, glass dropping 3 in., pkg of 12 (17 T 0230)	**A4, A17**
Probe and seeker, 5 1/8 in., each (14 T 0950)	**A22**
Safety goggles, SG34 regular, each (15 T 3046)	**A4, A14, A23**
Scales, bathroom, lb/kg, each (15 T 3800)	**A14**
Scissors, dissecting, each (14 T 0525)	**A3, A7, A8, A13, A20, A28**
Spring scale, pull-type, 5 kg/50 N, each (15 T 3775)	**A20**
Stereomicroscope, widefield, each (24 T 4601)	**A2, A22, A34**
Thermometer, lab −20 to 110°C, each (15 T 1416)	**A16, A34**
Thermometer, red alcohol, each (15 T 1460)	**A1, A4**
Tray w/cover 8 1/4 × 5 × 2 1/4 in., each (18 T 0300)	**A14**
Watch glass, syracuse, each (17 T 0530)	**A34**

Miscellaneous

Item	Lab
Apple (local)	A24
Blue cheese (local)	A2
Calculator (27 T 3055)	A3
Cardboard sheets (local)	A28
Egg, chicken (local)	A10
Egg, chicken, unfertilized (local)	A22
Flower pot, plastic 3 in., pkg of 10 (20 T 2130)	A20
Food product labels (local)	A5, A33
Garbage bag (local)	A14
Graph paper, 5 squares/in., pkg of 100 (15 T 3835)	A5, A29, A31
Ice cubes (local)	A4
Ice, crushed (local)	A24
Index cards, unlined 3 × 5 inches, pkg of 100 (15 T 9819)	A21
Light bulb, 150 W, 120 V Clear, each (36 T 4173)	A24
Marker, black lab, each (15 T 3083)	A7
Milk (local)	A16
Oil, vegetable, 16 oz (37 T 9540)	A1
Paper (local)	A6, A9, A11, A12, A15
Paper clips, pkg of 1000 (15 T 9815)	A7, A8
Paper towel, 100 sheet 2-ply roll, each (15 T 9844)	A25

Item	Lab
Paper, white (local)	A20
Pencils, colored, pkg of 12 (15 T 4690)	A1, A29
Plastic wrap, 12 in. wide, 50 ft roll (15 T 9858)	A1, A24
Poster board, white, 22 × 28 inches, pkg of 5 (15 T 9856)	A3
Pushpins, set of 5 colors, set (15 T 0505)	A7, A8
Ruler, 12 in. plastic, each (15 T 4655)	A28
Ruler, 6 in. white vinylite, each (14 T 0810)	A3, A7, A8, A13, A20
Spoon, plastic, each (250 T 8129)	A22
Spoons, plastic, pkg of 100 (15 T 9800)	A34
Stopwatch, digital, each (15 T 0512)	A4
Straws, plastic wrapped, pkg of 500 (15 T 9869)	A7, A8
String, .5 lb pkg (15 T 9869)	A20
Sugar, granulated, 5 lb pkg (39 T 3180)	A1
Swab applicator, pkg of 100 (14 T 5502)	A25
Tape, masking 3/4 in. × 60 yd. roll, pkg of 3 (15 T 9828)	A24
Tape, transparent w/dispenser, each (15 T 1959)	A3, A20
Teaspoon (local)	A16
Tote Box, polyethylene 21 × 12 × 6 in., each (18 T 9881)	A25
Yogurt (local)	A2, A16
Zipper bags, resealable 6 × 9 in., pkg of 10 (18 T 6922)	A24

Master Materials List, by Lab

A1 Imagining Solutions: Problem Solving

Required Materials	Quantity Needed
Beaker, low-form 400 mL Griffin	3 per team
Oil, vegetable	300 mL per team
Pen, black wax marker for glass	1 per team
Pencils, colored	3 per team
Plastic wrap, 12 in. wide	18 in. per team
Sugar, granulated	300 g per team
Thermometer, red alcohol	3 per team

A2 Comparing Living and Nonliving Things

Required Materials	Quantity Needed
Blue cheese	1 tsp per team
Coleus plant, living	1 per class
Cork, size 2	1 per class
Coverslip, 22 mm plastic	1 per team
Crustose lichen (SECT) QS Slide	1 per class
Earthworm (*Lumbricus*), living	1 per class
Elrathia kingi, large fossil	1 per class
Fantail goldfish, living	1 per class
Grasshopper, *Lubber,* preserved	1 per class
Microscope slide, precleaned	5 per team
Microscope	1 per team
Oriental mung bean seed	1 per team
Physarum polycephalum culture	1 per class
Pumice-Italy, SS	1 piece per class
Starch agar plate	1 plate per class
Stereomicroscope, widefield	1 per team
Yeast, viable	1 g per team
Yogurt	1 tsp per team

A3 Modeling Cells: Surface Area to Volume

Required Materials	Quantity Needed
Calculator	1 per team
Funnel, 58 mm polypropylene	1 per team
Graduated cylinder, glass 100 mL	1 per team
Poster board, white, 22 in. × 28 in.	1 sheet per team
Ruler, 6 in. white plastic	1 per team
Sand, fine white	5 g per team
Scissors, dissecting	1 per team
Tape, transparent	1 per team

A4 Demonstrating Diffusion

Required Materials	Quantity Needed
Apron, disposable polyethylene	1 per student
Beaker, low-form 400 mL Griffin	3 per team
Food coloring, 4 colors	15 drops per team
Gloves, heat defier kelnit cotton	1 per team
Hot plate, (700 W) single burner	1 per team
Ice cubes	2 per team
Pen, black wax marker for glass	1 per team
Pipet, glass dropping 3 in.	1 per team
Safety goggles, SG34 regular	1 pair per student
Stopwatch, digital	1 per team
Thermometer, red alcohol	1 per team

A5 Interpreting Labels: Stored Food Energy

Required Materials	Quantity Needed
Graph paper, 5 squares/in.	2 sheets per team
Food product labels	6 per team

A7 Making Models

Required Materials	Quantity Needed
Marker, black	1 per team
Paper clips	48 per team
Pushpins, set of 5 colors	48 per team
Ruler, 6 in. white plastic	1 per team
Scissors, dissecting	1 per team
Straws, plastic-wrapped	8 per team

A8 Making a Genetic Engineering Model

Required Materials	Quantity Needed
Paper clips	50 per team
Pushpins, set of 5 colors	56 per team
Ruler, 6 in. white plastic	1 per team
Scissors, dissecting	1 per team
Straw, plastic-wrapped	8 per team

A10 Analyzing Adaptations: Living on Land

Required Materials	Quantity Needed
Broad Windsor bean seed	2 per class
Chameleon, living American	1 per class
Egg, chicken	1 per class
Fantail goldfish, living	1 per class
Fern spores, hardy mixed	2 per class
Fern, living woodland	1 per class
Frog dev. state, 1-cell, pres.	2 eggs per class
Moss (*Polytrichum*), living	1 per class

A13 Using Random Sampling

Required Materials	Quantity Needed
Dish, specimen 8 1/4 in. × 3 1/4 in.	2 per team
Paper	1 piece per team
Ruler, 6 in. white plastic	1 per team
Scissors, dissecting	1 per team

A14 Determining the Amount of Refuse

Required Materials	Quantity Needed
Apron, disposable polyethylene	1 per student
Garbage bag	1 per team
Gloves, disposable medium	2 gloves per student
Safety goggles, SG34 regular	1 per student
Scales, bathroom, lb/kg	1 per team
Tray w/cover, 8 1/4 × 5 × 2 1/4 in.	1 per team

A16 Using Bacteria to Make Food

Required Materials	Quantity Needed
Aluminum foil, 12 in. wide roll	1 per team
Apron, disposable polyethylene	1 per team
Beaker, low-form 1000 mL Griffin	1 per team
Beaker, low-form 600 mL Griffin	1 per team
Gloves, disposable medium	2 gloves per student
Hot plate, (700 W) single burner	1 per team
Incubator, lab	1 per class
Milk	400 mL per team
Teaspoon	1 per team
Thermometer, lab −20 to 110°C	1 per team
Yogurt	1 per team
pH paper, 1-14 range	2 strips per team

A17 Observing Protists

Required Materials	Quantity Needed
Beaker, low-form 250 mL Griffin	1 per class
Coverslips, 22 mm plastic	2 per team
Euglena sp. culture	1 per class
Hot plate, (700 W) single burner	1 per class
Microscope slide, precleaned	2 per team
Microscope	1 per team
Paramecium caudatum culture	1 per class
Pipet, glass dropping 3 in.	2 droppers per team
Protist slowing agent: Detain	2 drops per team
Safranin 0	5 mL per class
Yeast, viable	1 packet per class

A18 Comparing Plant Adaptations

Required Materials	Quantity Needed
Begonia, living	1 per class
Ceratopteris, living	1 per class
Coleus plant, living	1 per class
Moss (*Polytrichum*), living	1 per class
Tradescantia, living	1 per class
Venus' flytrap, living	1 per class

A19 Inferring Function From Structure

Required Materials	Quantity Needed
Microscope	1 per team
Syringa (lilac) leaf (CS) QS Slide	1 per team

A20 Relating Root Structure to Function

Required Materials	Quantity Needed
Broad bean seed	3 per team
Flower pot, plastic 3 in.	3 per team
Paper, white	1 sheet per team
Ruler, 6 in. white plastic	1 per team
Scissors, dissecting	1 per team
Spring scale, pull-type, 5 kg/50 N	1 per team
String	1 ft per team
Sweet corn seed, untreated	3 per team
Tape, transparent	1 per team

A21 Recognizing Patterns of Symmetry

Required Materials	Quantity Needed
Animal survey set, preserved	1 per team
Apron, disposable polyethylene	1 per student
Dissection pan set	1 per team
Gloves, disposable medium	2 gloves per student
Index cards, unlined 3 × 5 in.	10 per team

A22 Comparing Animal Eggs

Required Materials	Quantity Needed
Dish, specimen 4 1/2 × 2 3/4 in.	2 per team
Egg, unfertilized chicken	1 per team
Forceps, medium dissecting	1 per team
Frog dev. state, 1-cell, pres.	1 egg per team
Magnifier, dual 3× and 6×	1 per team
Probe and seeker, 5 1/8 in.	1 per team
Spoon, plastic	1 per team
Stereomicroscope, widefield	1 per team

Suggested Substitute

Frog eggs (limited availability)	1 per team

A23 Observing Some Major Animal Groups

Required Materials	Quantity Needed
Animal phyla collection, preserved	1 per team
Apron, disposable polyethylene	1 per student
Dissection pan set	1 per team
Gloves, disposable medium	2 gloves per student
Safety goggles, SG34 regular	1 pair per student

A24 Observing Insect Behavior

Required Materials	Quantity Needed
Aluminum foil, 12 in. wide roll	1 sq ft per team
Apple	1 piece per team
Beaker, low-form 600 mL Griffin	2 per team
Crickets, living	1 cricket per team
Ice, crushed	1 per team
Lamp, with reflector and clamp	1 per team
Light bulb, 150 W, 120 V clear	1 per team
Plastic wrap, 12 in. wide	1 sq ft per team
Tape, masking 3/4 in. $\times$ 60 yd. roll	12 in. per team
Zipper bag, resealable 6 $\times$ 9 in.	2 per team

A25 Observing a Frog

Required Materials	Quantity Needed
Beaker, low-form 1000 mL Griffin	1 per team
Beaker, low-form 250 mL Griffin	1 per team
Grassfrogs, small living	1 per team
Paper towel, 100-sheet 2-ply roll	3 sheets per team
Swab applicator	1 swab per team
Tote box, polyethylene 21 $\times$ 12 $\times$ 6 in.	1 per team

A26 Vertebrate Skeletons

Required Materials	Quantity Needed
Grassfrog skeleton	1 per class
Perch skeleton, complete	1 per class
Pigeon skeleton, complete and mounted	1 per class
Skeleton, plastic rod mount	1 per class
Snake skeleton, complete and mounted	1 per class
Turtle skeleton, mounted	1 per class

A27 Comparing Skeletal Joints

Required Materials	Quantity Needed
Skull, plastic human	1 per team

A28 Bias and Experimentation

Required Materials	Quantity Needed
Cardboard sheets	six 3 $\times$ 5 cm pieces per team
Pins, nickel-plated T	11 per team
Ruler, 12 in. plastic	1 per team
Scissors, dissecting	1 per team

A29 Graphing Growth Rate Data

Required Materials	Quantity Needed
Graph paper, 5 squares/in.	1 sheet per team
Pencil, colored	2 per team

A31 Determining Lung Capacity

Required Materials	Quantity Needed
Balloons, 11 in. diameter	1 per student
Graph paper, 5 squares/in.	1 sheet per team
Meter stick	1 per team

Suggested Substitute

Lung volume bag set	1 per team

A32 Relating Cell Structure to Function

Required Materials	Quantity Needed
Human blood (SM) GS slide	1 per team
Microscope	1 per team

A33 Reading Labels: Nutritional Information

Required Materials	Quantity Needed
Food product labels	6 per team

A34 Culturing Frog Embryos

Required Materials	Quantity Needed
Apron, disposable polyethylene	1 per student
Beaker, low-form 100 mL Griffin	3 per team
Dish, specimen 4 1/2 $\times$ 2 3/4 in.	3 per team
Frog eggs (limited availability)	30 per team
Gloves, disposable medium	2 gloves per student
Hot Plate, (700 W) single-burner	1 per team
Pen, black wax marker for glass	1 per team
Pondwater	200 mL per team
Spoon, plastic	1 per team
Stereomicroscope, widefield	1 per team
Thermometer, Lab −20 to 110°C	1 per team
Watch glass	1 per team

HOLT BioSources
LAB PROGRAM

QUICK LABS

INCLUDES
LABS A1–A34

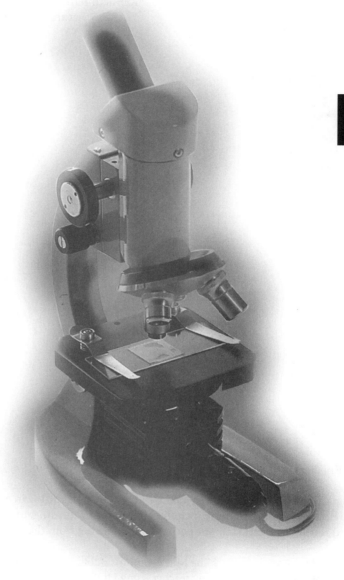

HOLT, RINEHART AND WINSTON
Harcourt Brace & Company

Austin • NewYork • Orlando • Atlanta • San Francisco • Boston • Dallas • Toronto • London

HOLT BIOSOURCES® *LAB PROGRAM*

Quick Labs

Acknowledgments

Copyediting

Amy Daniewicz
Denise Haney
Steve Oelenberger

Prepress

Rose Degollado

Manufacturing

Mike Roche

Design Development and Page Production

Morgan-Cain & Associates

Cover

Design—Morgan-Cain & Associates
Photography—Sam Dudgeon

Reviewers

Lab Activities
George Nassis
WARD'S Natural Science Establishment
Rochester, NY

Lab Safety
Kenneth Rainis
WARD'S Natural Science Establishment
Rochester, NY

Printed in the United States of America
ISBN 0-03-050693-X
1 2 3 4 5 6 7 8 022 00 99 98 97 96

QUICK LABS

BIOSOURCES
LAB PROGRAM

Contents

Contents continued

Name _____

Date _____ Class _____

HOLT
BIOSOURCES
LAB PROGRAM
QUICK LAB

A1 | *Imagining Solutions: Problem Solving*

This activity gives students an opportunity to use their imagination while comparing materials used to store solar energy. The activity can be extended, making it as open-ended as you wish. Allow the students to select the substances to be tested, but safety, cost, and availability should be considered.

Background

Imagination can be as important to the biologist as it is to a writer or story-teller. For this activity, you will use your imagination to suggest solutions to a problem involving heat storage. Problems such as the one you are about to encounter must be solved to make the use of solar energy practical.

Objectives

In this activity you will
• *imagine* solutions to a problem.

Materials

• heat-absorbing materials (3, your choice)
• beakers (3), 400 mL
• plastic wrap
• wax pencil
• thermometer °C
• colored pencils (3)

Preparation

Water, air, cooking oil, salt, sugar, starch, and gravel are materials to consider. Avoid materials that are toxic or inflammable.

If possible, have the students measure the temperatures of the beakers' contents first thing in the morning, during class, and at the end of the day. This work could be divided among members of the group.

1. Form cooperative groups of three students and consider the problem of heat storage. Propose three materials that could be used to collect and store heat. Take into consideration safety, availability, cost, and ease of use.
2. Have your teacher approve your proposed materials.
3. Make a table similar to the one shown below for recording your data.

Day	After 24 hours in the:	Temperatures of materials being tested		
1	Dark			
2	Sun			
3	Dark			

4. Fill each beaker with one of the materials to be tested. Cover each beaker with plastic wrap and label accordingly with a wax pencil.
5. Set the beakers in a dark place for 24 hours.
6. Remove the plastic wrap to measure the temperature of the material in each beaker, and record your results.
7. Re-cover the beakers with plastic wrap and transfer to a sunny location, such as the window sill, for 24 hours.
8. Remove the plastic wrap to measure the temperature of the material in each beaker, and record your results.

Procedure

Introduce the activity with a discussion as to why students think it is important to use the imagination. Guide students to develop a relationship between using the imagination to solve problems and technological advances.

9. Re-cover the beakers and return to the dark location for 24 hours.
10. Measure the temperature of the material in each beaker and record your results.
11. Use your colored pencils to make a line graph that compares the temperatures of the three beakers over time. Use a different color line to represent each of the beakers.
12. Clean up your materials and wash your hands before leaving the lab.

Analysis

1. **Summarizing Data** Summarize your findings.

Answers will vary depending on the materials being tested. The greatest fluctuation in temperature will probably occur in air.

2. **Analyzing Methods** At what point does imagination enter into the problem-solving process?

Imagination entered into the problem-solving process at the very beginning, when hypotheses are formulated.

3. **Evaluating Methods** Why might imagination be one of the most important traits of the biologist?

Imagination can lead to a hypothesis that can be tested. Without a hypothesis, the scientific process cannot continue.

Name _____

Date _____ Class _____

A2 Comparing Living and Nonliving Things

For many students, this may be their first exposure to "life" in a biological sense. They may associate life only with movement in animals and green color in plants. This activity should give the students an opportunity to explore a variety of life forms. The correct answer is not as important as showing evidence of observation and analysis. This activity is meant to initiate the learning cycle.

Background

Do all living things share certain traits? If so, what might these traits be? In this activity, you will try to answer these questions.

Objectives

In this activity you will
- **observe** and **compare** living and nonliving things.

Materials

See suggested specimen list on page 4.

- unlabeled specimens
- stereomicroscopes
- compound microscopes

Preparation

1. Form cooperative teams of two students.
2. Make a table similar to the one shown below.

Trait	Specimen									
	1	2	3	4	5	6	7	8	9	10
1.										
2.										
3.										
4.										
5.										
6.										
7.										
8.										
9.										
10.										

3. In the first column of the table, list 10 traits or processes that you associate with living organisms.
4. Closely observe a specimen. Follow any special directions for viewing the specimen that your teacher may supply.
5. Place an "X" in your data table for each trait you observe in the specimen.
6. Add to your table any new traits that come to mind during your observations.
7. Repeat steps 4–6 for each of the specimens.

Procedure

Let the students know that the emphasis of the activity is on observation and making comparisons. The traits identified at the beginning of the activity are a basis for future work.

Analysis

1. **Summarizing Data** Which specimens are alive? dead? nonliving?

Answers will vary according to the specimens observed.

2. **Comparing Groups** What traits do living things have in common?

Answers will vary. Some examples might include movement, eating, response to light or other stimuli, vocalization, excretion or elimination of waste, reproduction, growth, repair, or energy consumption.

3. **Comparing Groups** What traits do nonliving things have in common?

Answers will vary but generally are the converse of traits of living organisms.

4. **Evaluating Methods** Why can it be difficult to observe all the traits common to living things?

Answers will vary but should refer to temporal and perceptual limitations.

Place a variety of numbered specimens at different stations around the lab. Provide supplemental instructions whenever necessary. For example, instructions for using the microscope, moving the specimen, or removing a cover might be necessary. Select a wide variety of living, nonliving, and dead specimens. Specimens might include:

- unprocessed yogurt, include a slide for use under high power
- blue cheese or bread mold, include wet-mount slides
- dry mung beans or other seeds
- dried or activated yeast, show under high power
- pumice, clean sand, or powdered clay
- powdered charcoal, show under high power
- coral, sea shells, sponge, or snail shell
- dead insect
- living insect, crayfish, or earthworm
- goldfish or minnows

- cork, whole or thin-section under high power
- lichen on a rock, including a prepared slide under low power
- fossil
- loofa (Luffa) "sponge"—actually a dried plant stem available in health food stores
- synthetic sponge
- potted plant
- protozoans or algae collected from a stagnant pond or slow-moving stream
- slime mold growing on a moist paper towel in a petri dish or on plain agar in a petri dish

Name _____

Date _____ Class _____

A3 Modeling Cells: Surface Area to Volume

Surface area-to-volume relationships play a role in many aspects of biology. In this activity, the students build simple models to investigate why and how this ratio might limit the size of organisms.

Background

Are there limits as to how large organisms can grow? Some humans are very tall but do not grow as large as trees. Some large insects do exist but never grow to reach the sizes you might see in science fiction films. Why? One reason is that with increasing height there is a disproportionate increase in volume (or weight). If the height of an elephant were doubled, its weight would increase by eight times its original weight. An elephant cannot grow larger, because its legs could not support the increase in weight.

In this activity you will examine surface area-to-volume ratios on a small scale, using some model cells. You will use the collected data to reach some conclusions as to why this ratio might limit the size of a cell.

Objectives

In this activity you will
- **construct** and **analyze** various cell models.
- **measure** volume and calculate surface area.
- **calculate** surface-area-to-volume ratios.
- **form conclusions** about size limitations, using your data.

Materials

- scissors
- cell model cutouts (3)
- poster board
- tape
- metric ruler
- sand
- funnel
- large graduated cylinder
- calculator (optional)

Preparation

1. Form cooperative groups of two students. Cut out three cell models and fold each to form a three-dimensional shape. Cell dimensions should be recorded in the table in step 2 below. Use tape where directed so that the models hold their shape.

Procedure

2. Using the metric ruler, measure the length, width, and height dimensions of each model. Record the dimensions in a table like the one shown below.

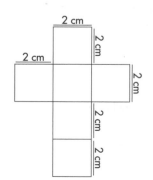

Cell	Dimensions (cm)	Surface Area (cm^2)	Volume (cm^3)	Ratio Surface Area to Volume
A	2 x 2	24	8	3.0
B	4 x 4	96	64	1.5
C	8 x 8	384	512	0.75

Graph paper can be used to construct the cell models. This facilitates drawing straight lines and right angles. If necessary, each team can be given predrawn patterns for each of the three cell models. Also, you might wish to select a range of measurements.

Have the class speculate on the factors that limit cells to microscopic sizes. Look for reasonable and logical responses rather than the correct answer. Discuss why models can be used to investigate some phenomena.

3. Fill each model with sand. Level off the sand at the top of the model, using the ruler.
4. Find the volume of sand in each model. You can do this two ways.
   a. Measure the amount of sand in each model, using a graduated cylinder. Pour the sand through the funnel into the graduated cylinder or a measuring cup.
      One millimeter = one cubic centimeter (1 cm³).
   b. Calculate the volume using the formula in step 2 of Analysis (below).

Analysis

1. **Analyzing Models** Complete the data table by calculating the area and volume of each model. To calculate total surface area for each model, find the area of each side (length × width) then multiply that number by 6. Enter the data in your table. Why do you need to multiply by 6?

   To calculate the total surface area, measures of the four sides, bottom, and top must be considered, making a total of six.

2. **Analyzing Models** To calculate the volume of sand, use the following formula:

   volume = length × width × height

   Record the volume of each model in your table.

3. **Analyzing Models** Calculate the surface area-to-volume ratio for each model. Use the following formula:

   $$\frac{\text{surface area}}{\text{volume}} = \text{ratio}$$

   Record the value in your table.

4. **Comparing and Contrasting Models** Which model has the largest surface area?

   Model C

   Which model has the largest volume?

   Model C

   Which model has the largest ratio?

   Model A

5. **Making Predictions** To maintain life, materials must be able to move into and out of a cell. What might be the advantage of having a large surface area?

   The cell could take in more materials.

   What might be the disadvantage of having a large volume?

   If the cell was larger, fewer of its interior structures would be near the cell membrane, making the cell less efficient.

Name _____

Date _____ Class _____

HOLT
BIOSOURCES
LAB PROGRAM
QUICK LAB

A4 Demonstrating Diffusion

While the demonstration of diffusion is the primary objective of this activity, the relationship between temperature and diffusion is used to rouse the student's curiosity. Although the diffusion of food coloring through a beaker of water is used as an example in many textbooks, many students have never actually seen it firsthand.

Background

Molecules must move about in a cell in order for a cell to survive. In this activity, you will see the movement of molecules and study how temperature affects this movement.

Objectives

In this activity you will
• **observe** diffusion.
• **recognize** the relationship between temperature and the rate of diffusion.

Materials

• beakers (3), 400 mL
• wax pencil
• water
• ice cubes
• hot plate

• insulated glove
• hot pad (optional)
• thermometer
• food coloring
• timer

Preparation

1. Form cooperative groups of two students. Complete steps 2–10.
2. Label three clean 400 mL beakers, *Cold*, *Medium*, and *Hot*, respectively.

Procedure

3. Add cool or lukewarm tap water to each beaker until it is approximately half full.
4. Place two ice cubes in the *Cold* beaker. Set both the *Cold* beaker and the *Medium* beaker aside.
5. **CAUTION: A hot plate's high temperature can cause injury.**
   Place the Hot beaker on a hot plate and heat the water to boiling.
6. **CAUTION: Boiling water can cause injury.**
   Use an insulated glove to carefully remove the beaker from the hot plate. Use the hot pad when placing hot objects on surfaces that are not suited for high temperatures.
7. Use a thermometer to find the initial temperature of the water in each beaker. Record your results in a table similar to the one shown.

Introduce the activity with a discussion of the importance of the movement of molecules into, out of, and within the cell. Define the term "operational definition" and explain that the students will form an operational definition of diffusion based on their observations. Their operational definitions can provide a basis for more detailed discussion later.

	Temperature of beaker		
	Cold	Medium	Hot
Initial temperature			
Time of diffusion			

If hot plates are not available, water can be boiled in a teapot or coffee-maker. Water can be chilled over-night in a refrigerator, if the use of ice cubes is inconvenient. A 9°C spread of temperatures from hot to cold should produce easily observed differences in the rate of diffusion.

8. **NOTE:** Do not shake or stir the water. Add five drops of food coloring to the beaker. Note the time required for the coloring to spread uniformly through the water. Record the result.
9. Remove the ice cubes from the *Cold* beaker. If necessary, add cold water to the beaker until its level is about equal to that of other beakers.
10. Repeat steps 7 and 8 for the *Cold* and *Medium* beakers.
11. Clean up your materials and wash your hands before leaving the lab.

Analysis

1. *Defining Terms* An operational definition is a definition based on observation. State an operational definition of *diffusion* based on your observations of the food coloring in the beakers.

Answers may vary. *Diffusion* should be defined in terms of the movement of molecules until they are uniformly distributed throughout the medium. Diffusion will always tend to produce uniform mixtures, since molecules move into areas where they are scarce.

2. *Analyzing Data* Explain how temperature affects diffusion.

Answers may vary according to the data but should indicate that the rate of diffusion is proportional to the ambient temperature.

3. *Making Inferences* Why is a warm body advantageous for a living thing?

Answers will vary. Accept any logical response that addresses the question. Insightful students should understand that warm body temperatures increase the rate of diffusion without any energy expenditure by the cell itself.

A5 Interpreting Labels: Stored Food Energy

In this activity, students compare the calorie contents of some common foods, using supplied data as well as the labels they have collected from other foods. Caloric values are standardized for 100 g samples and graphed to make comparison easier.

Background

Food supplies us with matter to build living tissue and energy to do work. The energy content of foods is measured in calories. How do foods differ in their energy content? In this activity you will interpret data from the labels of food products to answer this question.

Objectives

In this activity you will
- **graph** and **interpret** data pertaining to the energy content of various foods.

Materials

- food product labels (6)
- sheets of graph paper (2)

A few days before scheduling the activity, ask the students to form teams of two. Have each team collect the labels from at least six different food items.

Preparation

1. Form cooperative teams of two students. Complete steps 2–8.
2. Collect nutrition lists from six food products.
3. Make a table similar to the one shown below.

NOTE: You may substitute your choice of foods and calorie values.

Suggested Food List
Peanuts, 563 Cal/100 g
Ham, 95 Cal/100 g
Fudge cookies,
555 Cal/100 g
Crackers, 457 Cal/100 g
Corn, 71 Cal/100 g
Rice, 353 Cal/100 g

Food	Grams per serving	Calories per serving	Calories per gram	Calories per 100 grams

Procedure

Discuss the role of energy in living systems—ways in which humans use energy. Answer any questions students may have about the meaning of the terms. Some students may need assistance in converting ounces to grams and in making a bar graph.

Analysis

4. What types of nutritional information are included on your labels?

Answers will vary according to the foods sampled.

5. Select one label. Find the serving size and calories.

6. Calculate the number of calories per gram of food. Divide the serving size (in grams) by the number of calories per serving. If necessary, convert ounces to grams by multiplying the serving size by 28.4 g/oz.

7. Find the number of calories per 100 gram sample. Multiply calories per gram by 100. Record the results.

8. Repeat steps 5–7 for each of your labels.

9. Make a bar graph that compares the calorie content for 100-gram samples of each of the foods in your table.

1. *Analyzing Data* Which foods contain the most calories? The fewest calories?

Answers will vary according to the foods sampled. Of the foods initially listed in the table, peanuts and fudge cookies have the most calories, while corn and ham have the fewest calories.

2. *Evaluating Methods* Why is it helpful to convert the measurements to the same sample size?

Equal sample sizes make it easier to compare the calorie content for a given mass of food.

3. *Evaluating Methods* Why is calorie content on the label stated in terms of a serving size rather than in terms of a 100 gram sample?

The serving size on the label reflects the way most people should actually eat.

4. *Applying Concepts* How should a person put information about the calorie content to use?

People use nutritional information to plan a balanced diet.

Name _____

Date _____ Class _____

HOLT
BIOSOURCES
LAB PROGRAM
QUICK LAB

A6 Interpreting Information in a Pedigree

Schedule this activity prior to discussing the chapter or after a discussion of simple patterns of inheritance. The students construct and analyze pedigrees that summarize the family history of a trait or a set of traits in an easy-to-read diagram.

Background

Organizing information is often the key to solving a problem. Tracing the hereditary characteristics over many generations can be especially confusing unless the information is well organized. In this activity, you will learn how to organize hereditary information, making it much easier to analyze.

Objectives

In this activity you will
- **construct** and **analyze** a pedigree.

Materials

- paper
- pencil

Enhance this activity by providing examples of pedigrees. Medical genetics textbooks, the genetics department of a large hospital, and agricultural breeding services are some of the sources of such pedigrees.

Preparation

1. Pedigree I traces the dimples trait through three generations of a family. Blackened symbols represent people with dimples. Circles represent females and squares represent males.

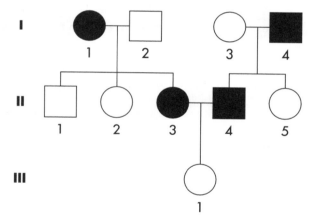

Pedigree I

Procedure

Introduce the activity by reading the hereditary information presented for one of the families in this activity. Have students describe the family without looking at their textbook. Their success should be minimal at best. Point out that the way information is organized can be crucial to using it and that this activity will show them a

2. The following passage describes the family shown in Pedigree I.

   Although Jane and Joe Smith have dimples, their daughter, Clarrisa, does not. Joe's dad has dimples, but his mother, and his sister, Grace, do not. Jane's dad, Mr. Renaldo, her brother, Jorge, and her sister, Emily, do not have dimples, but her mother does.

3. Write the name of each person below the correct symbol in Pedigree I. How are marriage and offspring symbolized? What do the Roman numerals symbolize?

   A straight line connecting a circle and square indicates a marriage, with a descending line leading to any offspring. Roman numerals identify each generation.

simple way to organize hereditary information.

Discuss the mechanics of Pedigree I before they continue with Pedigree II. Extend the activity by having the students construct pedigrees from summaries of novel situations, such as those involving maternal and fraternal twins, remarriage after divorce or death, and marriage across generations.

4. Make a pedigree based on the following passage about freckles.

Andy, Penny, and Delbert have freckles, but their mother, Mrs. Cummins, does not. Mrs. Giordano, Mrs. Cummins's sister, has freckles, but her parents, Mr. & Mrs. Lutz, do not. Deidra and Darlene Giordano are freckled, but their sister, Dixie, like her father, is not freckled.

Analysis

1. *Evaluating Techniques* What advantages does a pedigree have over a written passage?

Answers may vary but should indicate that a pedigree organizes hereditary information in less space, making it easier to read than the same information in a written passage.

2. *Summarizing Observations* How does a pedigree organize hereditary information, making it easier to understand?

Answers may vary but should indicate that a pedigree shows the number, gender, traits, and family relationship of each individual in a generation. All members of a generation are placed on the same line, identified by a Roman numeral. The symbols representing married individuals are joined by a straight line, with a perpendicular line descending to their offspring.

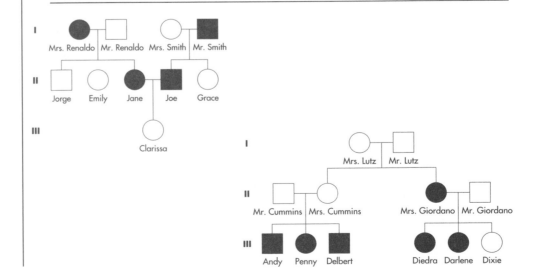

HOLT
BIOSOURCES
LAB PROGRAM
QUICK LAB

A7 *Making Models*

In this activity, students learn to construct and manipulate models. If scheduled before reading or discussing the chapter, the activity can provide an opportunity for exploration rather than confirmation. The students are guided through the construction of one model that they analyze, and then they use it as a basis for completing a second model. After students have completed the activity, the model can provide the basis for a detailed discussion of the structure of DNA.

Background

You know that DNA provides instructions that direct the activities in a cell. In this activity, you will build a model of DNA to help you understand its structure and functions in the cell.

Objectives

In this activity you will
• **construct** and **analyze** a model.

Materials

• plastic soda straws (8)
• centimeter ruler
• scissors
• permanent marker

• pushpins (12 of each color), red, blue, yellow, and green
• paper clips (48)

Preparation

1. Form cooperative teams of four students. Select one member of your group to form a team. Work with your teammate to complete steps 2–9.
2. **CAUTION: Pointed objects can cause injury if not properly used.** Cut the soda straws into 3 cm segments to make 48 segments.

Procedure

Discuss how models help us recognize patterns and relationships.

Have students cut straws into 3 cm segments prior to beginning the activity. Small, standard-sized paper clips work well. Colored pushpins are available through office supply stores.

3. Insert a pushpin midway along the length of each straw segment. Push paper clip into one end of each straw segment until it touches the pin.
4. Keep the pins in a straight line, and insert the paper clip of a blue-pushpin segment into the open end of a red-pin segment. Add segments of straw to the red-segment end in the following order: green, yellow, blue, yellow, blue, blue, green, red, blue, and green. Use the permanent marker to label the blue segment on the end "top." This strand of segments is one-half of your first model.

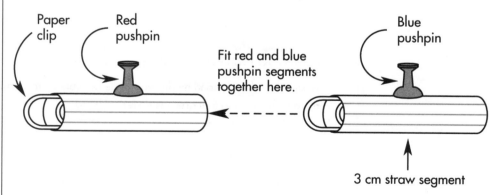

Paper clip   Red pushpin

Fit red and blue pushpin segments together here.

Blue pushpin

3 cm straw segment

5. Begin to construct the other half of your first model with a yellow-pin segment. Keep the pins in a straight line. Link segments together in the following order: green, red, blue, yellow, blue, yellow, yellow, red, green, yellow, and red. Label the yellow segment on the end "top."

HRW material copyrighted under notice appearing earlier in this work. ·

6. Place the strands parallel to each other on the table with the "top" blue pin of one strand facing the "top" yellow pin of the second strand. What color pin is always across from a blue pin?

Yellow is always across from blue.

What color pin is always across from a red pin?

Green is always across from red.

7. Use 12 of the remaining straw segments with pins of any color, in any order, to make one-half of another model.

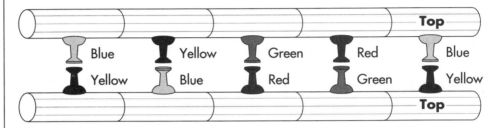

8. Exchange your team's 12 remaining straw segments and the strand that your team made in step 7 with those of the other team in your group.

9. Using the straw segments and the strand from the other team, make a strand of segments that has the correct sequence of color pins to complete your second model.

Analysis

1. **Comparing and Contrasting Models** How are your two models similar?

Answers may vary but should indicate that each model has pins of four different colors, such that blue faces yellow and red faces green.

How do they differ?

The models differ in how these colors are sequenced on a strand.

2. **Analyzing Models** What did you learn about the first model that enabled you to construct a second model?

Construction of the second model was made possible by recognizing that yellow is always paired with blue and that green is paired with red .

Name _____

Date _____ Class _____

HOLT
BIOSOURCES
LAB PROGRAM
QUICK LAB

A8 *Making a Genetic Engineering Model*

This activity gives students the opportunity to use models similar to those made in Quick Lab A7 to investigate processes relating to genetic engineering. This is meant to be an introductory activity that provides students with a concrete example of a sophisticated abstract process.

Background

In this activity, you will use simple models to demonstrate the manipulation of genetic material to produce new combinations of traits.

Objectives

In this activity you will
- **construct** a model.
- **demonstrate** processes related to genetic engineering.

Materials

- soda straws, 3 cm pieces (50)
- centimeter ruler
- scissors
- pushpins (14 of each color), red, blue, yellow, and green
- paper clips (50)

Preparation

1. Form cooperative groups of four students. One 2-person team should complete steps 2 and 3 while the other team completes steps 4–6. Work with your entire group to complete steps 7–10.

Procedure

Small, standard-sized paper clips work well in this model. Colored pushpins are available through office supply stores. The students can cut straws into 3-cm segments at home or prior to beginning the activity.

2. **CAUTION: Pointed objects can cause injury if not properly used.** Make a model of a bacterial DNA molecule similar to the one described in Quick Lab A7. Arrange the nucleotides of the master strand in the following order: blue, red, green, yellow, red, red, blue, blue, green, red, blue, green, red, blue, blue, green, yellow, and red.

3. With your double-stranded DNA model lying on the table, form a circular molecule by carefully joining the opposite ends of each strand. Make a sketch of the molecule that shows the arrangement of the bases.

Review the nature of the models to be used and the structure of DNA before beginning the activity. Do not attempt to introduce new vocabulary at this time.

4. **CAUTION: Pointed objects can cause injury if not properly used.** Make one strand of a donor, human DNA molecule similar to the one described in Quick Lab A7. Use the following sequence: blue, blue, red, red, yellow, green, green, blue, red, and yellow.

5. Make a complementary strand of donor DNA that has the following sequence: blue, red, red, yellow, green, blue, yellow, yellow, green, and green.

6. Match the complementary portions of the two strands of your human DNA fragment. What is unusual about the structure of this donor DNA fragment?

Opposite strands at each end of the DNA fragment have four unpaired, exposed nucleotides.

Make a sketch of the donor molecule.

BBRRYGGBRY
   BRRYGBYYGG

7. Imagine that an enzyme moves around the circular molecule of bacterial DNA until it finds the sequence red-red-blue-blue and its complementary sequence green-green-yellow-yellow. Find this sequence in the bacterial molecule that you drew in Step 3.

8. Simulate the action of the enzyme by splitting the circular molecule at the sequence you identified in Step 7. Separate the yellow nucleotide from the blue at one end of the sequence, and the green from the red at the opposite end of the sequence on the complementary strand. Make a sketch of the split molecule.

9. Move the double-stranded donor, human DNA fragment into the break in the bacterial DNA molecule.

10. Imagine that a second enzyme joins the ends of the donor and recipient DNA creating a new DNA molecule. Make a sketch of the final bacterial DNA molecule.

Analysis

1. **Making Comparisons** How does the original bacterial DNA molecule differ from the final DNA molecule?

   The new bacterial molecule has more nucleotides than the original molecule as a result of the addition of the human DNA fragment.

2. **Applying Concepts** Of what possible benefit could this process be to humans?

   Answers may vary. A human gene could be inserted into a bacterial DNA molecule.

Name _____

Date _____ Class _____

A9 Comparing Observations of Body Parts

This activity gives the students the opportunity to make and analyze observations relating to homologous structures. Use it as the introduction to the chapter, or after the discussion of the evidence of evolution.

Background

Could you tell if two strangers were related just by looking at them? What kinds of evidence would help you determine their relationship? In this activity you will observe parts of various animals and look for evidence that these animals are related to one another.

Objectives

In this activity you will
- *describe* structures of different organisms.
- *identify* relationships between the structures of different organisms.

Materials

- paper
- pencil

No additional materials are required. Skeletons of vertebrate limbs would enhance the activity.

Preparation

1. Make a table to record your observations of the limbs of seven different animals. Include columns across for the limb's *shape*, the *number of bones in the upper limb*, the *number of bones in the lower limb*, a description of the *arrangement of the bones*, and the limb's *function*. List the names of the animals down the left side of the table.

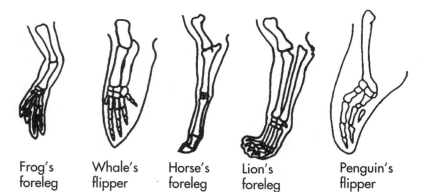

Frog's foreleg   Whale's flipper   Horse's foreleg   Lion's foreleg   Penguin's flipper

Procedure

Stress that biologists often must rely on observations made from the drawings of others, at least during the initial stages of an investigation. Explain that in this activity, students must make the best use of the drawings of appendages.

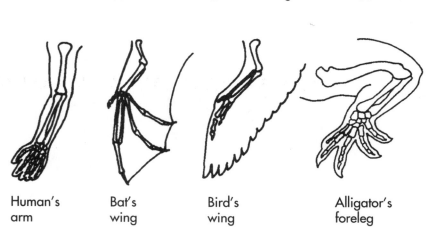

Human's arm   Bat's wing   Bird's wing   Alligator's foreleg

Analysis

2. Record your observations in your table.

1. *Analyzing Observations* How are the limbs of the frog, whale, lion, human, bat, bird, and alligator similar?

Answers may vary but should indicate that the number and arrangement of bones in the upper and lower limbs of the vertebrate animals are similar.

How do they differ?

The shape and function may differ.

2. *Analyzing Observations* Which limbs perform similar functions?

Appendages of the bird and bat are used for flight. The human limb is used for grasping and manipulation of objects, the whale's and penguin's in swimming, and the others' in locomotion on land.

3. *Identifying Relationships* Which is the better indicator of the relationship between two organisms—structure or function? Explain your reasoning.

Answers may vary but should suggest that structure is a better indicator of relationship than is function.

A10 Analyzing Adaptations: Living on Land

This activity gives students an opportunity to observe some of the adaptations that allowed life to move out of the water. As students compare the structures of various pairs of aquatic and terrestrial organisms, they should observe the need to avoid desiccation and the need to acquire oxygen—important factors in moving to a terrestrial environment.

Background

The move from a watery environment to the land was a giant step in evolution. In this activity you will study some of the adaptations that allowed organisms to make the move to land.

Objectives

In this activity you will
- **observe** various structural features.
- **relate** structural features to the move to a terrestrial environment.

Materials

- various aquatic and terrestrial specimens

Preparation

Place pairs of specimens around the laboratory for inspection. Each pair should consist of one terrestrial specimen and one aquatic specimen showing an adaptation for life on land. Include the name of each organism and any additional directions needed to draw your students' attention to the related adaptation.

1. Form cooperative teams of two students. Complete steps 2–4.
2. Make a table similar to the one shown below for recording your observations.

Specimen pairs		Observations	Adaptions for life on land
Name	Habitat		
1.	Aquatic		
	Terrestrial		
2.	Aquatic		
	Terrestrial		
3.	Aquatic		
	Terrestrial		
4.	Aquatic		
	Terrestrial		

Procedure

Discuss the differences the students perceive between terrestrial and aquatic environments.

3. Pairs of specimens are displayed around the classroom. One member of each pair is identified as an organism that lives in the water while the other lives on land. Observe one pair of specimens. Record their names and your observations.
4. Review your observations and suggest reasons why one organism is better adapted for life on land than the other.

Analysis

1. **Summarizing Data** List the features of land organisms that make them better adapted for life on land.

   Answers may vary according to the specimens observed but might include roots, a waxy covering on plants, a waterproof exoskeleton, a seed with a seed coat and stored food, and lungs.

2. *Analyzing Data* Based on your observations, describe two factors that organisms had to overcome to survive out of water.

Accept any logical answer. Desiccation and the need to obtain oxygen are the two most

important obstacles to overcome in order to survive on land.

Pairs of specimens could include any or all of the following:
• a chicken's egg and a frog's egg to compare shelled to nonshelled eggs
• an insect and a flatworm, jellyfish, or sponge to show the cuticle or the exoskeleton of the insect
• a seed and a spore to show the seed coat and food supply
• organisms with lungs and those that use diffusion or gills to obtain oxygen
• a bryophyte and a plant with an extensive root system
Another pair of specimens can include a fresh apple and an apple that has been dipped briefly in gasoline to remove the cutin. Allow both apples to sit for two days before this activity. The apple without the cutin should be shriveled.

Name _____

Date _____ Class _____

HOLT BIOSOURCES
LAB PROGRAM
QUICK LAB

A11 Comparing Primate Features

Structural similarity is one indicator of evolutionary relationship. In this activity, students compare the skull, jaw, and spinal attachment, and hand of the gorilla with those of the human. The similarities should suggest the existence of a common ancestor.

Background

Paleontologists are scientists who look for and study fossils. Structural similarities can indicate an evolutionary relationship between organisms. In this activity you will observe structural features of primate hands, jaws, and skulls.

Objectives

In this activity you will
• *observe* primate structures.
• *relate* these features to evolutionary relationships.

Materials

• paper
• pencil

No materials are required.

Preparation

1. Form cooperative teams of two students. Complete steps 2–7.
2. Make a table similar to the one shown below for recording your observations.

	Comparison of human and ape features	
	Human	Gorilla
Hand		
Jaw		
Skull		
Spinal attachment		

Procedure

Review the sources of evidence of evolutionary relationship presented earlier.

3. Compare the structure of the gorilla's hand below to your own hand. Try to touch the tip of your thumb to the tip of every other finger on that hand. Does it appear that a gorilla could do this?

Human hand   Gorilla hand

4. Record your observations of the similarities between these hands.

5. Compare the jaws of the human and the gorilla. Describe the similarities and differences.

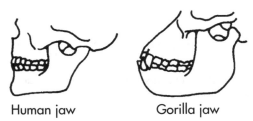

Human jaw   Gorilla jaw

6. Compare the structures of the human skull and the gorilla skull. Describe the similarities and differences.

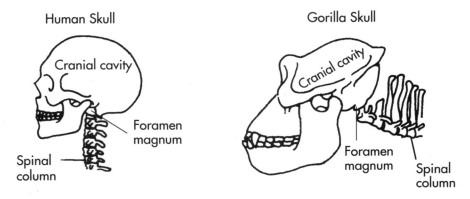

Human Skull   Gorilla Skull

7. Use the figure above to compare how the spinal column attaches to the skulls of the human and the gorilla. Describe the similarities and differences you observe.

Analysis

1. *Summarizing Observations* Summarize the evidence suggesting that apes and humans may have had a common ancestor.

Answers may vary. Students should describe such similarities as the opposable thumb, the number and shapes of teeth, and the general shape and structure of the skull. Differences include the shape and size of the jaw, the attachment point of the spinal column to the skull, the size of certain teeth, and the number of teeth.

2. *Making Inferences* How does the structure of the hand contribute to the ability to use tools?

An opposable thumb allows primates to grasp tools more tightly.

3. *Identifying Relationships* How does skull structure provide evidence of a primate's relative intelligence?

Answers may vary. Students should suggest that the size of the portion of the cranium that surrounds the brain indicates the size of the brain. This factor, in turn, may provide evidence for inferences about an organism's relative intelligence.

Name _____

Date _____ Class _____

A12 *Making a Food Web*

This activity introduces students to the concepts of food chains and food webs. The activity draws on everyday experience and common sense rather than on detailed knowledge of a particular ecosystem.

Background

One organism consumes another for energy and raw materials. A food chain shows the sequence in which energy passes from one organism to another as it moves through the community.

Objectives

In this activity you will
- **categorize** organisms.
- **make** a food web.

Materials

- paper
- pencil

No materials are required. Slides, transparencies, or photographs showing organisms common to a woodland habitat can be helpful.

Preparation

1. Form cooperative groups of two students to complete this activity.
2. Closely observe the illustration on this page showing a portion of a community. List all the organisms that you see.

A woodland community

3. Add to your list other organisms that might also be present in this community, but are not shown.

<u>Answers may vary. Students should list the organisms shown in the illustration and</u>
<u>additional organisms likely to be present in a woodland habitat.</u>

Procedure

Discuss the types of organisms that might inhabit the community and be depicted in the activity. Feel free to substitute a local ecosystem for the one shown or to extend the activity by using an ecosystem more familiar to your students.

Analysis

4. On a separate sheet of paper, write the name of one organism from your list that is capable of photosynthesis.
5. Draw a short arrow leading from this organism to the name of a second organism that might eat it.
6. You have drawn the first two links of a food chain. Extend your chain to three links by adding an arrow and a third organism that might consume the second.
7. Extend your food chain to five links.
8. Make two more food chains consisting of five links each.
9. Construct a food web by adding arrows that connect organisms in different chains. Make as many connections as possible.

1. **Comparing Data** How are food chains similar to food webs, yet different?

   A food web is composed of individual food chains. It includes links between food chains that in themselves form additional food chains.

2. **Evaluating Methods** Why can a food web be more helpful than a description of the same information written in a paragraph?

   A food web presents the data in a graphical representation, making it easier to see relationships among the various links.

Name _____

Date _____ Class _____

A13 *Using Random Sampling*

In this activity students compare data obtained from an estimate by sampling with data obtained from an actual count. Emphasize that estimating by sampling can yield reasonably accurate population data. Have the students conduct an actual field study or come up with techniques to reduce the percentage error in their estimates.

Background

Scientists cannot possibly count every organism in a population. One way to estimate the size of a population is to collect data by taking random samples. In this activity you will look at how data obtained by random sampling compare with data obtained by an actual count. Sampling is used to track population growth in an ecosystem. It is one of many methods used by scientists to collect data when studying ecosystems.

Objectives

In this activity you will
- *estimate* by sampling.
- *count* the number of sunflower plants in a meadow.
- *compare* gathered data.
- *calculate* percentage error.

Materials

- unlined paper
- ruler
- pencil
- containers (2)
- scissors

Preparation

1. Cut a sheet of paper into 20 slips, each approximately 4 cm × 4 cm.
2. Number 10 of the slips from 1 to 10, respectively. Put the numbered slips in a small container.
3. Label the remaining 10 slips from A through J and put them in a second container.
4. To record your data, make a table similar to the one shown.

Method	Average number of plants per grid segment	Total number of plants
Sample		
Actual count		

Procedure

For those flowers that fall on the grid lines, have students include plants that fall on the line at the top or the right of each square. Ignore those that fall on the line at the bottom or the left of a square.

5. The grid shown here represents a meadow measuring 10 m on each side. Each grid segment is 1 m × 1 m. Each black circle represents one sunflower plant.

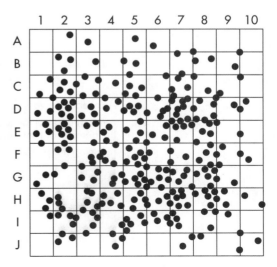

6. Randomly remove one slip from each container. On a separate sheet of paper, write down the number-letter combination you drew and find the grid segment that matches that combination. Count the number of sunflower plants in that combination. Count the number of sunflower plants in that grid segment. Record this number on the separate sheet of paper. Return each slip to the appropriate container.

7. Repeat step 6 until you have data for 10 different grid segments. These 10 grid segments represent a sample. Gathering data from a randomly selected sample of a larger area or group is called *sampling*.

8. Find the total number of sunflower plants for the 10-segment sample. This is an estimate, based on sampling, of the number of plants in the meadow. Divide this number by 10 to determine the average number of sunflower plants per square meter in the sample. Record this number in the table. Multiply the average number of sunflower plants per square meter by 100 to find the total number of plants in the meadow. Record this number in your table.

9. Count all the sunflower plants actually shown in the meadow. Record this number in your table. Divide this figure by 100 to calculate the average number of sunflower plants per square meter. Record your data.

Analysis

1. ***Comparing Data*** Compare the actual count with the data you recorded for your random sampling estimate.

Answers will vary according to the data collected. There are 261 sunflower plants actually represented in the meadow.

2. ***Analyzing Methods*** Why was the paper-slip method used to select grid segments?

Answers may vary but should suggest that the paper-slip method is a way to select the segments randomly.

3. ***Evaluating Methods*** What is the percentage error in your estimate? To calculate percentage error find the difference between the actual count and the estimated count. Divide this difference by the actual count and multiply by 100.

$$\frac{\text{Difference}}{\text{Actual count}} \times 100 = \% \text{Error}$$

Answers may vary, but low percentage error should suggest that estimating by sampling can provide data that are reasonably close to actual counts.

4. ***Making Inferences*** How could you change the procedure in this activity to reduce your percentage error?

Answers may vary, but students should recognize that making sure that the sample is truly random and not localized in any way reduces percentage error. Also, increasing the number of random samples provides data that are closer to the actual counts, thereby reducing the percentage error.

Name _____

Date _____ Class _____

A14 *Determining The Amount of Refuse*

This activity gives students firsthand knowledge of recycling and the problems associated with generated refuse.

Background

How much refuse does a family produce every day? How much of this material could be recycled? In this activity you will collect data to answer these questions.

Objectives

In this activity you will
* **infer**, **measure**, and **compare** the amount of refuse your family produces each day

Materials

* disposable gloves
* apron
* goggles
* plastic garbage bags
* bathroom scale or other weighing device (at home)

Preparation

Each student should have access to a bathroom scale for measuring the mass of refuse produced. If this is not possible, give the students small plastic refuse bags. Have the students measure in "refuse bag" units. Afterward, you can help them convert to a standard volume unit or to mass using a conversion factor based on the average mass in a typical bag of refuse.

1. How many kilograms of refuse do you think your family produces each day? (NOTE: Each pound is equal to 0.454 kg.)
2. Make a table similar to the one shown below.

Day	Type of refuse		
	Recyclable	Reusable	Waste
1			
2			
3			
4			
5			
6			
7			
Average			
Total (yr)			

Procedure

3. **CAUTION: Wear disposable gloves, an apron and goggles while completing the following activity.** Use fresh refuse from within your own home only; do not work with refuse more than one day old or with refuse from outside your home.

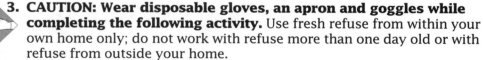

4. Sort through your waste and garbage containers at home. Separate the contents into three groups: materials that could be recycled, materials that could be reused, and materials that are truly waste.
5. List the items in each group.
6. Place the items in each group in a plastic garbage bag.

7. Weigh each bag on a bathroom scale. Multiply the weight in pounds by 0.454 to obtain the weight in kilograms. Record your data. NOTE: If you do not have the use of a bathroom scale, see your teacher for special instructions.
8. Repeat steps 3–7 each day for 7 days.
9. Calculate the average for each group of materials and record your answer in your table.
10. Calculate the total amount of yearly refuse produced by your family by multiplying the sum of the averages of the three groups of refuse by 365. Record your results in your table.
11. Clean up your materials and wash your hands before leaving the lab.

Analysis

1. **Summarizing Data** Summarize your data.

   Answers will vary.

2. **Analyzing Data** How would recycling affect the total amount of refuse produced by your family?

   Answers will vary but should indicate that recycling would significantly reduce the amount of refuse produced.

3. **Making Inferences** Why would re-use and recycling programs benefit our society?

   Answers may vary but should indicate that these programs should benefit all of society by placing less demand on resources and the environment, as well as adding to the economic base.

Have students prepare a list naming the types of things that can be recycled. Remind them that nonaluminum cans and plastic bottles can also be recycled. You might also have them locate the local recycling facility.

Direct the students to collect their data each day from the refuse within their own home. Caution them not to work with refuse that is more than one day old because of health hazards.

Name _____

Date _____ Class _____

HOLT
BIOSOURCES
LAB PROGRAM
QUICK LAB

A15 | *Grouping Things You Use Daily*

The process of classification is introduced in this activity by having students consider a system for grouping their own possessions. Classification systems are encountered every day; food stores, department stores, recorded-music stores, and warehouses are just a few of the places that commonly employ a system of classification.

Background

Classification systems are part of our everyday lives. You find things placed in groups in food stores, in record stores, and even in your home. In this activity, you will explore how you group things you use daily.

Objectives

In this activity you will
• **identify** types of things that you group.
• **explain** why you group things.

Materials

• paper
• pencil

Preparation

1. Make a table similar to the one shown below.

Level		
I	II	III
A. Clothing	1.	a. b.
	2.	a. b.
	3.	a. b.
	4.	a. b.
B. School	1.	a. b.
	2.	a. b.
	3.	a. b.
	4.	a. b.
C. Recreation	1.	a. b.
	2.	a. b.
	3.	a. b.
	4.	a. b.

2. Close your eyes and picture your possessions. Think of all the things that you use in everyday life.
3. List your possessions. Be as specific as possible.

Procedure

Emphasize the process of classification and its application in everyday life. Some students might find it easier to make their initial list of possessions while at home. Students can share their characteristics, but avoid having them share their lists of possessions.

Analysis

4. Divide your list into three smaller lists—clothing, school-related items, and recreational items.
5. Divide each list into four subgroups. Use logical characteristics as the basis for your subgroups. For example, casual clothing or books might be two subgroups.
6. Record the characteristics used for these subgroups in the Level II column of your table.
7. List the items that belong to each subgroup.
8. Where possible, divide each subgroup into two groups. Use logical characteristics as a basis for these groups and record this information in Level III of your table.
9. List the items that belong to each of these groups.

1. **Analyzing Observations** Do the characteristics used to distinguish the various groups become more general or more specific as you move from Level I to Level III?

   They become more specific.

2. **Evaluating Techniques** Explain how you might use a grouping or classification system of your possessions.

   Answers will vary but should suggest that systems are often grouped to make them easier to find.

3. **Applying Concepts** Explain how classification systems are used by merchants or other businesses.

   Answers may vary but should suggest that merchants group items so shoppers can find them more easily.

A16 *Using Bacteria to Make Food*

While students and the general public are very much aware of the harmful effects of bacterial contamination, they are less familiar with the beneficial uses of microorganisms. In this activity, students make a simple yogurt culture. With reasonable attention to cleanliness and good lab technique, the yogurt can be tasted after its production.

Background

Although bacteria can cause disease and destroy other forms of life, they also have many beneficial uses. In this activity, you will have the opportunity to use bacteria to make a food product.

Objectives

In this activity you will
• **observe** the production of yogurt.

Review the concept of pH and acids and bases. Demonstrate the use of pH test strips if the students are unfamiliar with them.

Materials

• apron
• disposable gloves
• milk, 400 mL
• beaker, 600 mL
• pH test strips
• hot plate

• beaker, 1000 mL
• water
• thermometer
• yogurt (plain, with active cultures)
• teaspoon
• aluminum foil

Preparation

1. Form cooperative teams of four students. Complete steps 2–8.
2. What is pH and how is it measured?

Procedure

Commercial yogurt can be used as a starter culture. Active yogurt cultures commonly contain *Streptococcus thermophiles* and *Lactobacillus bulgaricus.*

Only clean glassware should be used. The inoculated milk cultures should be incubated at about 39°C.

3. **CAUTION: Put on a laboratory apron and disposable gloves.** Leave them on throughout this activity.
4. To avoid contamination, keep your work area and equipment as clean as possible. Clean all glassware, spoons, and thermometers thoroughly before beginning this activity and upon its completion.
5. Pour 400 mL of milk into a clean 600 mL beaker.
6. Measure the pH of the milk with a pH test strip. Record the results.
7. **CAUTION: The hot plate produces high temperatures that can cause a serious burn.** Place the beaker of milk in a 1000 mL water bath. Heat the milk to 81°C for 15 minutes. Be careful to avoid boiling the milk.

8. Allow the milk to cool to about 39°C.
9. Add one teaspoon of yogurt to the milk and stir gently until mixed. Cover the beaker with aluminum foil.
10. Incubate the milk at about 39°C for 24 hours.
11. Measure the pH of the newly formed yogurt. Record your results.

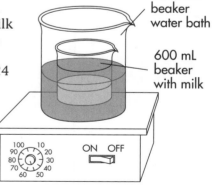

1000 mL beaker water bath

600 mL beaker with milk

The pH of a solution is the concentration of the hydronium ions expressed as the negative of the common logarithm. It is measured on a scale that ranges from 0 to 14, describing the acidity or alkalinity of a solution.

Analysis

1. **Summarizing Observations** Describe the changes that occurred in the milk.

   Answers may vary but should describe a change in consistency and a change in pH in the culture. The milk's initial pH of about 7.0 will decrease as the bacterial action makes the culture slightly acidic.

2. **Analyzing Data** What evidence do you have that a chemical reaction occurred in the milk?

   Answers may vary but might describe the change in consistency and the change in pH as evidence that a chemical reaction has occurred.

3. **Applying Concepts** What other evidence could be collected to indicate that yogurt was produced in the beaker?

   Answers may vary but might suggest that a microscopic comparison of the new culture and the starter culture could provide evidence that yogurt was produced. Also, a taste and smell test could provide additional evidence.

Name _____

Date _____ Class _____

HOLT
BioSources
LAB PROGRAM
QUICK LAB

A17 *Observing Protists*

This investigation gives students an opportunity to observe cultures of identified protists. Discuss techniques that can improve observations. Have students explore why drawing and taking notes can improve their powers of observation. Discuss the benefits of listing items to look for, and the benefits of searching for relationships among observations.

Background

A glass of water taken from a pond can contain thousands of microorganisms. In this activity you will have the opportunity to view some examples of these organisms.

Objectives

In this activity you will
• *observe* and *compare* characteristics among protists.

Materials

• medicine droppers
• methyl cellulose
• coverslips
• yeast-Safranin O solution
• *Paramecium* culture
• glass slides
• compound light microscope
• additional protist cultures

Preparation

Obtain a variety of protists from WARD'S. *Paramecium, Blepharisma, Amoeba, Euglena, Stentor, Vorti-cella,* and *Volvox* are just a few suitable genera. Select genera that exhibit a variety of structural characteristics, methods of locomotion, and types of feeding responses.

Purchase 1% Safranin O solution from WARD'S for staining the yeast cells.

1. Name the five kingdoms and give an example of each.

_____

_____

_____

How can you distinguish among the organisms of each kingdom?

_____

_____

_____

_____

2. Make a table similar to the one shown below to record your observations of protists.

Protist name	Sketch	General description	Type of movement	Feeding mechanism

Procedure

3. Review the procedure for making a wet mount.
4. Form cooperative teams of two. Complete steps 6–8.
5. Place a drop of methyl cellulose on a clean glass slide to slow the movement of the protists. Place a drop of *Paramecium* culture on the methyl cellulose. Gently add a coverslip.
6. Use the low power setting on your microscope to locate the organisms on the slide. Switch to high power and focus on one *Paramecium* for several minutes. Record your observations in your table.
7. In your table, make a labeled drawing of this organism. Include the organism's name and the magnification that the drawing represents. Write a brief description of your observations next to your drawing.
8. Using a clean dropper for each culture, repeat steps 4 through 7 using each of the other cultures. Record your observations in your table.
9. Clean up your materials and wash your hands before leaving the lab.

Analysis

1. *Analyzing Observations* What characteristics are common to all of the organisms that you observed?

   Answers may vary but should indicate that all the specimens were unicellular and possessed a nucleus bounded by a nuclear membrane.

2. *Analyzing Observations* In what ways do these organisms differ?

   Answers will vary according to the specimens selected but should discuss such characteristics as size, color, feeding response, and method of locomotion.

3. *Analyzing Observations* What evidence do you have that some protists might be capable of photosynthesis?

   Some specimens are green, suggesting the presence of chlorophyll.

4. *Inferring Conclusions* How might a protist's ability to move relate to its method of obtaining nutrition?

   Answers should exhibit an understanding that locomotion allows the specimen to search for food. Protists that are sessile must wait for food to approach or be capable of making their own food.

5. *Identifying Relationships* Based on observable characteristics, make a simple biological key that could be used to identify the organisms you observed.

   Answers will vary, but students should be able to use such characteristics as color, method of locomotion, and feeding response to develop a key.

Name _____

Date _____ Class _____

A18 Comparing Plant Adaptations

This activity gives students the opportunity to observe a variety of plant adaptations. Emphasize observation rather than naming plant structures. Based on their observations, students are asked to draw inferences about the evolutionary advantages of these adaptations.

Background

Plants are found in almost every habitat on Earth. They exhibit a variety of sizes, shapes, and structures. In this activity you will observe a variety of plants.

Objectives

In this activity you will
- *observe* and *compare* vascular and nonvascular plants.

Materials

- a number of plant specimens

Preparation

1. Form cooperative teams of two students. Complete steps 3–10.
2. Use a separate sheet of paper to prepare the following table for recording your observations. Across the top, on the first line, center and write the heading: *Characteristics*. On the second line write the following headings: *Size in cm; Multicellular Structure; Habitat; Roots, Stems,* or *Leaves; Flowers; Seeds; Evidence of Internal Tubes; Waxy Covering on Leaves; Openings in Surface of Leaf;* and *Division of Labor.* Down the left side, write the number and name of the specimen.

Procedure

Discuss the nature of the traits to be observed in the lab and the evidence that would indicate the existence of each trait in a particular specimen. Review any procedures to be used that are unique to the specific specimens being used.

3. Observe all the demonstration materials at your lab station. These may include a living plant, a microscopic preparation, or graphic materials.
4. Based on your observations of the materials, record information about size, multicellular structure, habitat, and the presence of roots, stems, leaves, flowers, and seeds in your table.
5. Observe the surface of the living specimen and the surface of the microscopic preparation, if one is available, for evidence of internal tubes. Record your observations.
6. Observe the surfaces of the specimens, if possible, for evidence of a waxy covering. Record your observations.
7. If available, observe the microscopic preparation of a leaf. Look for evidence of openings in the surface. Record your observations.
8. Is there evidence of a division of labor in this plant? Record your answer in your table.
9. Repeat steps 3–8 for each of the plants assigned by your teacher.

Analysis

1. **Summarizing Observations** Summarize your observations of the plants.

   Answers will vary according to the specimens observed.

   _____

2. **Identifying Relationships** What traits seem to be common to all plants that have roots, stems, or leaves?

   Evidence of internal tubes, multicellularity, and division of labor are common to all plants with roots, stems, and leaves.

   _____

3. **Making Inferences** What traits appear to be common to all plants living in a dry habitat?

Evidence of a waxy covering, seeds, internal tubes, multicellularity, and division of labor are common to plants from a dry habitat.

4. **Analyzing Data** Describe the traits of a highly evolved plant and explain your reasoning.

Answers will vary but should describe all the traits observed in this activity as being present in a highly evolved plant.

Make a variety of specimens available for observation. Include live specimens, microscopic preparations, and graphic materials as necessary. Provide clues to direct the students' attention to structural features you feel they may otherwise miss. Include specimens of algae, mosses, liverworts, ferns, gymnosperms, and angiosperms.

Name _____

Date _____ Class _____

HOLT
BIOSOURCES
LAB PROGRAM
QUICK LAB

A19 *Inferring Function From Structure*

This activity gives students the opportunity to explore the structure of a leaf and to make simple inferences based on their observations. Develop interest in the chapter rather than introducing the details or terminology of leaf anatomy.

Background

A leaf is a complex structure composed of a variety of tissues working together to help the plant survive. The primary function of a leaf is food production. In this activity, you will explore the structure of a leaf and make inferences about leaf function based on your observations.

Objectives

In this activity you will
• *infer* function from *observations* of leaf structure.

Materials

• lilac leaf cross section, prepared slide
• microscope

Prepared slides of lilac leaf cross sections can be obtained from WARD'S (91 M 8720).

Preparation

1. Form cooperative teams of two students. Complete steps 2–7.
2. List the functions of vascular tissue in plants.

   Vascular tissue in plants transports water, food, and minerals throughout the plant.

3. List the specific tasks a leaf must perform to carry out its primary function.

   Specific tasks should correspond with photosynthesis.

Procedure

Review the process of photosynthesis and the general structure of a vascular plant. Have students list what they know of the functions performed by the roots, stems, and leaves.

4. View a prepared slide of a lilac leaf cross section under low power. Find an area of the leaf that looks similar to the drawing below. If necessary, move the slide to observe the entire length of the section.
5. Observe this section of the leaf under high power.
6. Make a drawing showing the structure of this section of the lilac leaf.
7. Describe this section of the leaf in detail. Pay particular attention to the areas labeled A–F below, but base your description on your specimen.

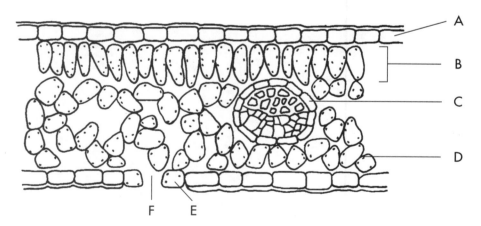

Lilac leaf cross section

Analysis

1. **Using Relationships** How do the location and structure of area A in the leaf provide evidence of its function?

   Answers may vary but should suggest that area A covers the inner tissues and provides protection.

2. **Making Inferences** What does the shape of structure C suggest about its possible function?

   The round, tubelike shape suggests that area C functions as tubes to transport materials to and from the leaf.

3. **Making Inferences** What do the contents of the cells in area B suggest about their function?

   The cells in area B contain green structures, suggesting that photosynthesis occurs in this area. Also this area's location just beneath the upper surface provides the most light, as well as protection from the outer environment.

4. **Identifying Relationships** How might carbon dioxide enter the leaf? Explain your reasoning.

   Answers may vary but should suggest that carbon dioxide could enter through opening F of the drawing, in the bottom of the leaf.

HOLT
BIOSOURCES
LAB PROGRAM
QUICK LAB

A20 Relating Root Structure to Function

This activity gives students the opportunity to investigate the relationship between the root structure and the soil-anchoring ability of two food plants. The loss of soil to water and wind erosion depletes the amount of land available for growing food products. Plants reduce the amount of soil lost to erosion.

Background

Erosion depletes the amount of land available for growing food products. Plants that provide food while binding the soil are of special value to us. In this activity, you will investigate the relationship between the root's structure and its ability to hold the soil.

Objectives

In this activity you will
- *analyze* data.
- *evaluate* inferences.

Materials

- paper
- lima bean seedlings (3)
- corn seedlings (3)
- adhesive tape
- scissors
- cord
- hand-held spring scale
- centimeter ruler

Preparation

1. Form cooperative groups of four. Work with one member of your group to complete steps 3–11. One pair in the group will use lima bean seedlings while the other pair uses corn seedlings.
2. On a separate sheet of paper make a table similar to the one shown below.

Seedling No.	Lima bean seedlings			Corn seedlings		
	Force N	Length mm	Diameter mm	Force N	Length mm	Diameter mm
1						
2						
3						
Average						

Procedure

Demonstrate how to read the scale. Caution students to pull the plants with a steady motion. A snapping motion is more likely to break the stem and would be very difficult to measure.

3. Select three pots of the same type of seedling. Place a small piece of adhesive tape around each stem at the level of the soil.
4. Attach one end of a cord to the tape and the other end to a spring scale.
5. Carefully, while reading the indicator, pull the scale upward and slightly away from the plant, as shown in the drawing below.
6. As the plant is uprooted, note the force on the indicator. Record the results.
7. Place the plant on a piece of paper.
8. Measure the length and diameter of the root system. Record your results.
9. Repeat steps 4–8 for each plant.
10. Find the average force, length, and diameter for your sample.
11. Combine your data with that of the other team in your group.
12. Use the averages to make a bar graph that compares the measurements for both kinds of plant.

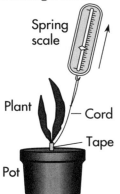

Analysis

1. **Summarizing Data** Summarize the data in your graphs.

Answers may vary but should adequately summarize the data.

2. **Identifying Relationships** Describe any evidence of a relationship between the force required to pull the plant from the soil and the other measurements.

Answers will vary according to the data collected. There will likely be a relationship between the force required to pull the plant from the soil and the length and diameter of the root system.

3. **Making Predictions** Which root system would be most effective in an area where wind erosion could be a problem? Explain your answer.

Plants such as corn and other grasses have fibrous roots that bind soil effectively, making them suitable for areas where erosion can be a problem.

4. **Identifying Relationships** Why is knowledge of plant structure helpful to a farmer when choosing a crop's location?

Answers may vary but should suggest that a knowledge of plant structure would help a farmer make wise decisions about the location of crops.

A21 Recognizing Patterns of Symmetry

This activity introduces students to the concept of symmetry. They observe an assortment of animals and group them based on body plan. More formal definitions of radial and bilateral symmetry can be developed later. At this stage, develop the concept of symmetry.

Background

Symmetry is the regular arrangement of parts around a point on an object or body. In other words, symmetry can be thought of as a plan for the arrangement of body parts. In this activity you will try to recognize the types of symmetry found in the animal kingdom.

Objectives

In this activity you will
• **observe** animals.
• **identify** different types of symmetry.

Materials

• index cards (one per specimen), 3 × 5 in.
• an assortment of animals

Preparation
Procedure

1. Form cooperative teams of two students. Complete steps 2 through 6.

2. Closely observe one animal.
3. Write the animal's name on an index card.
4. Draw a simple sketch of the animal showing its external appearance.
5. Write a brief description of the animal below its sketch.
6. Repeat steps 2–5 for each animal you observe.
7. Sort your cards into three groups—A, B, and C—based on the symmetry of the animals.

Analysis

Discuss the definition of symmetry presented in the activity. Have students give examples of symmetry or lack of symmetry in objects they encounter in everyday life. Do not introduce the terms radial and bilateral at this time.

1. **Summarizing Observations** Describe the traits common to each of your groups of animals.

   Answers will vary depending on the specimens used but should relate to the body plans
   observed.

2. **Identifying Relationships** The word "radial" means extending outward from a center point. Why can this word be used to describe one type of symmetry in animals?

   Answers will vary but should explain how animals such as the sea star, jellyfish, and sea
   urchin have parts that radiate from a central point on the body.

Have students move from station to station to observe an assortment of animals showing asymmetry, and radial and bilateral symmetries. Specimens might include the following: a number of different sponges, a jellyfish, a sea urchin, a sea star, a snail, a grasshopper, a crayfish, a frog, a lizard, a gerbil, or other choices. Use live specimens whenever possible. Have students supply their own index cards. A variety of specimen may be purchased from WARD'S.

3. *Identifying Relationships* The word "bilateral" means having two sides or halves. Why can this word be used to describe one type of symmetry in animals?

Answers will vary but should explain how animals such as the crayfish, grasshopper, and

gerbil have parts that form two sides that appear to be mirror images of each other.

4. *Making Inferences* Sponges are generally considered to be asymmetrical. What must this description mean?

Answers may vary but should suggest that the word "asymmetry" means the lack of an

organized body plan.

Name _____

Date _____ Class _____

HOLT
BIOSOURCES
LAB PROGRAM
QUICK LAB

A22 Comparing Animal Eggs

In this activity, students compare the egg of an animal that develops on land to the egg of an animal that develops in the water. The relative proportion of food stored in the egg and the existence of a hard shell are two adaptations that should be easily observed.

Background

To survive in the harsh environment of the land, animals evolved adaptations relating to reproduction and development. Some of these adaptations can be observed in the structure of their eggs.

Objectives

In this activity you will
- **compare** and **contrast** eggs of land and water animals.

Materials

- small spoon
- preserved frog eggs
- culture dish
- water
- hand lens or stereomicroscope
- blunt probe
- forceps
- chicken eggs

Preparation

Review the requirements for maintaining life. Caution students to treat the frog eggs gently.

1. Form cooperative teams of two students. Complete steps 2–16. Use a separate sheet of paper to make a table for recording your observations and drawings.
2. What does an embryo require to develop normally?

   food, water, protection, a means of gas exchange, and waste removal

   _____

3. What functions does an egg perform?

   The egg provides a suitable environment for the developing embryo until it is able to exist independently.

   _____

Procedure

4. Use a small spoon to transfer a frog egg to a clean culture dish. Add enough water to cover the egg.
5. Observe the egg using a hand lens or the highest magnification on your stereomicroscope. Record your observations.
6. Use a probe to gently touch the egg's surface. Record your observations.
7. Make a drawing of the frog egg. Show the relative proportion of the egg's light and dark material.
8. In general, the darker portion of the egg will develop into the embryo. The lighter portion of the egg will provide food for the developing cells. What fraction of the egg provides food for these living cells?

   Answers will vary depending upon the amount of the lighter portion which decreases during embryonic development.

   _____

Order three to four eggs per team, to allow for damage. The eggs can be reused, but expect some damage. Chicken eggs can be obtained from a grocery store. Fertile eggs are not required for this activity.

9. Return the frog egg to its storage container.
10. Observe the exterior of a chicken egg. Record your observations.

HRW material copyrighted under notice appearing earlier in this work.

11. Add tap water to a clean culture dish until it is about one-quarter full. Gently crack the eggshell on the edge of the dish and separate the shell, allowing the contents to flow into the water.
12. Examine the rounded end of the eggshell's exterior and interior surfaces. Record your observations.
13. Use your forceps to examine the small white disk on the yolk's surface. This structure contains the living egg cell that would develop into the embryo. The remainder of the egg provides food and water for these cells. What fraction of the egg provides food and water for the living cells?

Answers will vary. Students should suggest the majority of the egg.

14. Draw the contents of the egg as it appears in the dish.
15. Write a detailed description of the contents of the egg.
16. Clean up your materials and wash your hands before leaving the lab.

Analysis

1. **Making Comparisons** Compare and contrast the frog and chicken eggs.

Students should discuss the absence or presence of a shell, relative size, and relative proportion of food in each egg.

2. **Identifying Relationships** Describe the structure of the eggshell and explain its purpose.

Students should discuss the presence of a thin membrane along the inner surface and the hard exterior covering of the eggshell, which protect the embryo from foreign material and drying. The students should suggest that since the frog develops in water, a shell is not required.

Why is a shell necessary for the survival of the chicken embryo but not for the frog?

Oxygen must be able to move into the egg, while carbon dioxide must move out.

3. **Identifying Relationships** What factors might explain why a chicken egg contains so much more food and water than the frog egg?

Answers may vary but should suggest that more food may be required for the chicken because it is larger and its time of development is much longer than that of the frog.

HOLT
BioSources

LAB PROGRAM

QUICK LAB

A23 Observing Some Major Animal Groups

In this activity the students are asked to observe a small group of diverse specimens based on a list of traits of their own choosing. The focus of this activity should be on their choice of traits rather than on the specimens themselves.

Background

Since science is based on observation, knowing what to observe is very important to your eventual success. In this activity, you will observe a variety of animals and identify the traits you believe to be most meaningful.

Objectives

In this activity you will
- **observe** some major groups of animals.
- **evaluate** traits in terms of their importance in describing animals.

Materials

- one specimen each of a number of different kinds of animals

Preparation

Each team of two students should have a minimum of six specimens to observe. While the students can move from station to station observing single specimens, having all six at one station will facilitate comparisons. The specimens used by each team can vary, although a diverse group is recommended. Try to include one specimen from each phylum. Refer to WARD'S catalog for a variety of specimens.

The Preparation questions point the students in the direction of two important traits used to compare animal groups: highest level of organization and symmetry. Review these concepts before continuing with the activity. Allow students to select their own list of traits to observe so they can discover that some traits may be more meaningful than others.

1. Form cooperative groups of four students. Work in pairs to complete steps 2–6.
2. List and define the levels of organization exhibited by living things, from the lowest to the highest.
3. Describe and give an example of each type of symmetry exhibited by animals.
4. Make a table similar to the one shown below. **NOTE:** Your headings may differ from those on the sample table below.

Trait	Animal specimen					
	Sponge	Jellyfish	Flatworm	Earthworm	Crayfish	Frog
Highest level of organization						
Symmetry						

Procedure

5. Observe the animal specimens.
6. List and describe six traits that you can use to describe these animals. For example, each animal can be described in terms of the highest level of organization it exhibits.
7. Share your list of traits with the other team in your group. Discuss the traits and decide on the best four to use to describe animals. Record these four traits in the first column of your table.
8. Observe your assigned animals in terms of the traits in your table. Record your observations.

Analysis

1. **Summarizing Observations** Describe, in general terms, how these animals are similar and how they differ.

   Answers will vary but should discuss the traits of the six specimens.

2. **Evaluating Methods** Which trait or traits would you say are most important?

   Answers will vary but should suggest that some traits may have a bearing on others.

   Explain your reasons.

   For example, the level of organization is more significant than the existence of a specific organ system.

3. **Evaluating Methods** Which trait or traits would you say are less important in describing animals?

   Accept any logical answer.

   Explain your reasons.

4. **Applying Concepts** Based on your experiences, what factors are important in describing animals?

   Answers will vary but should be logical. Traits that affect the existence of other traits are the most significant.

   Explain your reasons.

   These traits are generally of greater evolutionary significance because they have such a major bearing on the adaptive advantage of the organism.

HOLT
BIOSOURCES
LAB PROGRAM
QUICK LAB

A24 *Observing Insect Behavior*

This activity gives students the opportunity to observe simple adaptive behaviors in crickets. Use this activity to stimulate interest for further study of insect structure and behavior.

Background

With more than 750,000 species, insects may be the most successful group of animals on Earth. This activity allows you to observe some aspects of structure and behavior that help make insects so successful.

Objectives

In this activity you will
- *observe* cricket behavior.
- *locate* and *identify* common external structures.

Materials

- cricket
- beakers (2), 500 mL
- plastic wrap
- apple
- masking tape
- aluminum foil
- self-locking freezer bags (2)
- crushed ice
- hot tap water

Preparation

1. Form cooperative groups of four students. Work with one member of your group to complete steps 2–12. Use a separate sheet of paper for recording your observations.

Procedure

2. **CAUTION: Use humane treatment when handling live specimens.** Place a cricket into a clean, 500-mL beaker and quickly cover the beaker with plastic wrap. The supply of oxygen in the beaker is sufficient for you to complete your work.
3. Add a small piece of apple to the beaker. Set the beaker on the table. Sit quietly for several minutes and observe the cricket. Any movement will cause the cricket to stop what it is doing. Record your observations.
4. Remove the plastic wrap from the beaker and quickly attach a second beaker. Join the two beakers together, at the mouths, with masking tape.
5. Wrap one beaker with aluminum foil.
6. Gently tap the sides of the beaker until the cricket is uncovered. Lay the joined beakers on their sides with a bright lamp over the uncovered beaker.
7. Without disturbing the cricket, carefully move the aluminum foil to the other beaker. Observe the location of the cricket after five minutes. Record your observations.
8. Fill a plastic freezer bag half full with crushed ice. Fill another bag half full with hot tap water. Seal each bag and arrange them side by side on the table.

Caution students to gently handle the crickets. Discuss the appropriate methods for collecting and transporting the animals while in the lab.

9. Gently rock the joined beakers from side to side until the cricket is in the center. Place the joined beakers on the freezer bags as shown below.

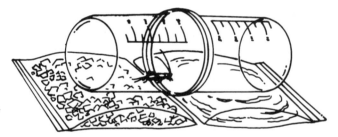

10. Observe the cricket's behavior for five minutes. Then record your observations.
11. Perform steps 9 and 10 two more times. Then record your observations.
12. Compare your observations with those of the other team in your group. Discuss the possible reasons for any differences.
13. Clean up your materials and wash your hands before leaving the lab.

Analysis

1. **Making Inferences** Explain why giving crickets small dishes filled with sugar solution would not keep them alive.

   Answers will vary but should suggest that the cricket's mouth is designed for chewing
   rather than lapping or sucking.

2. **Identifying Relationships** What evidence suggests that crickets are sensitive to light?

   Answers may vary but should describe the movement of the cricket from areas of bright
   light to darkness.

3. **Making Inferences** How does the cricket's response to temperature improve its chance of survival?

   Answers may vary but should relate the cricket's need for adequate warmth to its response
   to temperature.

A25 Observing a Frog

In this activity the students make simple observations of the external anatomy of a common amphibian—the frog. While most students have seen frogs, many have not had an opportunity to study a live frog closely. Approach the activity as an exploration that can provide a concrete basis for further detailed discussion.

Background

Amphibians can be found in a variety of habitats. You have probably seen a live frog or toad before, but perhaps you have never closely observed one. In this activity, you will observe the features of a common amphibian—the frog.

Objectives

In this activity you will
- **observe** amphibians.
- **describe** characteristics.

Materials

- large plastic tub
- paper towels
- beaker, 250 mL
- water

- frog
- beaker, 1000 mL
- gauze
- cotton swab

Preparation

Procedure

Caution the students about the proper handling of the specimens. The frogs need to be kept moist to prevent injury. Ask students to close their eyes and picture a frog before they actually view the specimens. Have them write a description of the image they see. Emphasize that this activity focuses on observations; they should not hurry, thinking that they have already seen all there is to see.

1. Based only on past experience, describe in detail the structure and general appearance of a frog. Make a table for recording your observations and drawings.
2. Form cooperative teams of three. Complete steps 3–13.
3. Cover the bottom of the tub with three layers of paper towels. Fill the 250 mL beaker with tap water for wetting the towels. (**NOTE:** Keep water on hand to keep towels from drying.)
4. **CAUTION: Use humane treatment when handling live specimens.** Use a 1000 mL beaker to carry a frog to your lab table.
5. Place the frog on the wet towels in the center of the tub. Cover the frog with the beaker. Wait 3–5 minutes while the frog calms. Avoid moving or touching the glass.
6. Slowly remove the beaker and observe the frog. Make a detailed drawing showing its external features. Label the head, abdomen, and appendages.
7. Observe the frog's head. Look closely at its eyes and look for evidence of ears. Record your observations by adding to your drawing.
8. Watch the frog as it breathes. Record your observations.
9. Look closely at the frog's forelimbs and hind limbs. Observe its fingers, feet, and toes. Record your observations in additional drawings.
10. Run your fingertip along the frog's skin. Compare the way it feels with the way your skin feels. Record your observations.
11. Gently prod the frog with the tip of a moistened cotton swab. Observe its movement. Pay special attention to the way the individual parts of its legs bend. Show your observations in a sketch of the frog's movement.

12. Return the frog to the beaker and then to the aquarium.
13. Clean up your materials and wash your hands before leaving the lab.

Analysis

1. **Summarizing Observations** Summarize your observations by writing a detailed description of a frog's anatomy.

   Answers will vary but should provide a more detailed description of the frog than that presented in the preparation.

2. **Making Comparisons** How is a frog similar to you?

   Similarities might include symmetry, the number of eyes, and the movement and number of appendages.

   How is a frog different from you?

   Answers will vary. Students could discuss differences associated with the attachment of the head to the body, in the number of fingers, and the parts of the leg.

Name _____

Date _____ Class _____

HOLT
BioSources
LAB PROGRAM
QUICK LAB

A26 | *Vertebrate Skeletons*

This activity gives students the opportunity to observe vertebrate skeletons and to make inferences about evolutionary relationships. Skeletal features provide easily observable evidence of homologies that form the basis for these inferences.

Background

Skeletal features can reveal relationships among animals. Similarities provide evidence that different animals evolved from a common ancestor. Bones also provide evidence of ecological relationships that exist among animals. In this activity, you will observe a number of vertebrate skeletons and make inferences based on your observations.

Objectives

In this activity you will
- *compare* vertebrate skeletons.
- *infer relationships* based on homologies.

Materials

- a variety of skeletal specimens or other resources

Preparation

The activity can be completed using photographs and drawings of skeletons, however, actual specimens (a perch, a pigeon, a garter snake, a turtle, a cat, a frog, and a human skeleton) are recommended. One specimen of each can be shared by the entire class.

1. Form cooperative teams of two students. Complete steps 2–10. Use a separate sheet of paper for recording your observations.
2. What are homologous structures?

    A homologous structure is a body part with the same basic structure as that of another
    organism, suggesting common ancestory.

3. Describe the type of evidence that might suggest that your hand, a pigeon's wing, and a cat's paw are homologous.

    Answers may vary but should suggest that the number, arrangement, and general shape of
    the bones that make up the human hand, pigeon wing, and cat paw are sources of
    evidence of homology.

Procedure

Have students move around the lab stations, recording their observations of various specimens. They can complete the activity at their desks, returning to a specimen if necessary.

4. Examine the available models and illustrations of vertebrate skeletons below.

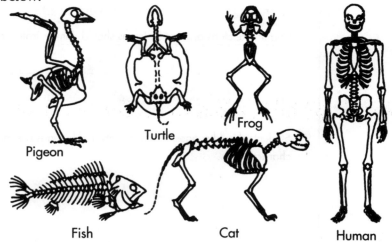

Pigeon Turtle Frog Human

Fish Cat Human

5. List four similarities and three differences that you observe.
6. Examine the backbone of each skeleton and describe their similarities and their differences.
7. Which two animals have backbones that are most alike?

   With exception of the tail, the cat and the human.

8. A human upper limb includes the shoulder, upper arm, forearm, wrist, and hand. How many bones make up each segment?

   Three bones make up the shoulder, one bone the upper arm, two bones the forearm, eight bones the wrist, and twenty bones the hand.

9. Observe the bones of the wrist and hand. Record your observations.
10. Compare the human upper limb with the forelimbs of the cat, the pigeon, the turtle, and the frog. Record your observations.

Analysis

1. *Analyzing Observations* Which animal's forelimb is most like the human upper limb?

   The human upper limb and cat forelimb are most alike.

   Describe the observations that led you to this conclusion.

   The bones are similar in number, arrangement, and relative size.

2. *Analyzing Observations* What evidence do you have to support the inference that vertebrates evolved from a common ancestor?

   Much homology is observed in the skeletons of vertebrates. Many similarities in the numbers, arrangements, and relative sizes of the bones of various structures support the hypothesis that vertebrates evolved from a common ancestor.

3. *Analyzing Observations* Based on your observations in this lab, why should humans and cats be considered more closely related than humans and any of the other vertebrates you observed?

   Human skeletons and cat skeletons should demonstrate the greatest degree of homology.

   How is the classification of humans and cats consistent with this conclusion?

   Students should suggest that both cats and humans are mammals sharing many characteristics.

Name _____

Date _____ Class _____

A27 | *Comparing Skeletal Joints*

Students explore five types of skeletal joints in this activity. They observe and analyze the movements of the major types of skeletal joints. Next, they conduct a survey of other joints in the body, identifying each according to its type.

Background

The bones of your skeleton come together in *joints*. In this activity, you will explore the types of skeletal joints found in your body by observing an example of each type of joint and surveying the other joints in your body.

Objectives

In this activity you will
• *identify* and *compare* skeletal joints.

Materials

• paper
• pen
• human skull or other animal skull

Have a real or artificial skull to illustrate immovable or fixed joints. The skull of any animal can be substituted for a human skull. A detailed illustration will suffice, if necessary.

Preparation

1. Use a separate sheet of paper to make a table for recording your observations of the skeletal joints shown below.

Pivot joint

Hinge joint

Ball-and-socket joint

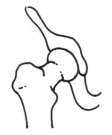

Gliding joint

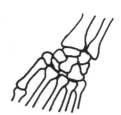

Fixed cranial joint

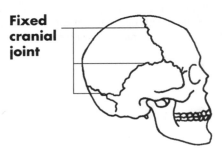

Procedure

Tell the students to use common sense and care while making the movements described in the activity. Only normal movements are required. Review the procedure for examining the movement of the pivot joint. Holding the neck is necessary to illustrate the limited side-to-side movement of this joint.

2. **CAUTION: Do not move in ways that can cause injury or pain.** Straighten your index finger as if to point. Slowly bend the finger until it touches your palm. Moving only your finger, try to move the individual parts of the finger in other directions. Record your description of the movements in your table.

3. Move your arm in as many ways as possible. Record your description of the movements first from the shoulder, then from the elbow.

4. Place your hands on the sides of your neck to hold the neck in place. *Gently* move only your head in all possible directions. Try not to let your neck bend. Record your observations of the movements.

5. Use your left hand to grip your right arm just above the wrist. Without moving your forearm, move your hand in all possible directions. Record your observations of the movements.

6. Examine the joints in the top of a skull. Gently press on each side of the joints to see if movement is possible without damaging the skull. Record your observations.

7. Starting with your feet, examine the movement of each of the other joints in your skeleton. Record the other locations in the body where you discover each type of joint.

Analysis

1. *Comparing Observations* Rank the five types of joints according to their freedom of movement. Start with the joint that allows the least freedom of movement.

Fixed cranial joint, pivot joint, glide joint, hinge joint, and ball-and-socket joint

2. *Analyzing Observations* Which types of joints are involved in walking?

Ball-and-socket joint, hinge joint, and glide joints

3. *Applying Concepts* For each type of skeletal joint, name a common, nonliving object that has a similar type of joint in its construction.

Answers might include the expansion joint in bridges, a beacon, small disc or ball attached to the underside of furniture legs to allow easy sliding, door hinges, ball-and-socket wrenches, and lamps.

Procedure Answers

2. Hinge joints allow the index finger to move in a forward and backward direction.
3. A ball-and-socket joint allows the wide movement between the arm and shoulder, whereas the elbow is a hinge joint with limited movement.
4. Rotating movement by a pivot joint allows the movement between the head and the neck.
5. Gliding joints allow movement of the wrist.
6. Fixed cranial joints allow no movement.

Name _____

Date _____ Class _____

HOLT
BIOSOURCES
LAB PROGRAM
QUICK LAB

A28 | *Bias and Experimentation*

Touch relies on two types of receptors, one type located deep in the dermis, the other located just below the epidermis. This activity allows students to determine the distribution of these touch receptors, using a simple procedure that reduces the amount of error caused by the expectations of the subject.

Background

A biased answer is based on your beliefs rather than on observation. The possibility of bias exists when a person's evaluation is required in an experiment. For example, you see an object touching the skin but do not actually feel it. Because you expect to feel it, you might actually say that you feel the object. Your answer is biased by your expectations. In this activity, you will try to reduce the effect of such bias in determining the distribution of touch receptors in the skin.

Objectives

In this activity you will
• *measure* the distance between touch receptors.
• *understand* the need to *reduce bias* in experimentation.

Materials

• cardboard
• scissors
• dissecting pins (11)
• metric ruler

Preparation

Use clean dissecting pins and caution the students to only gently touch the pins to the skin's surface. Thick corrugated cardboard holds the pins upright, keeping their tips the measured distance apart.

1. Form cooperative teams of two students.
2. Cut cardboard into six 3×5 cm rectangles.
3. Make a *touch-tester* by inserting two pins, 2 mm apart, halfway through one piece of cardboard, as shown below.

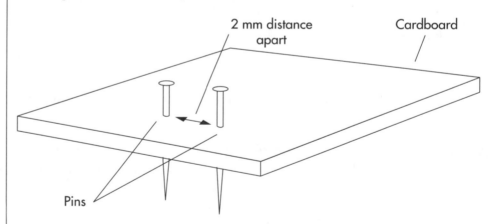

2 mm distance apart   Cardboard

Pins

4. Place the other pairs of pins 5 mm, 1 cm, 2 cm, and 3 cm apart in the other pieces of cardboard. Push the last pin through the center of the last rectangle. Each touch-tester is used to touch the tips of the pins to the skin.
5. When two pins are so close together that they stimulate only one touch receptor, a person feels only one pin. *How could these touch-testers be used to find the distance between touch receptors?*

Answers will vary but should suggest a method in which the minimum distance is found when two objects touched to the skin can be sensed as two objects.

6. Record your data in a table similar to the table below.

Distance between pins	Number of pins perceived when touched					
	Back of neck	Back of hand	Fingertip	Inside forearm	Outside forearm	Lips
2 mm						
5 mm						
1 cm						
2 cm						
3 cm						

Procedure

Discuss the procedure for determining the distribution of touch receptors.

Later in the activity, have the class discuss techniques that could be employed to reduce this type of error. Explain how double-blind techniques are used to reduce the effect of bias.

CAUTION: Test subject must close eyes and remain motionless. Tester must apply the pins gently. Avoid piercing the skin.

7. Use the touch-tester to gently touch both pins to the back of your partner's neck. Record a "+" if two pins are perceived and a "–" if one pin is felt. Repeat the test for three trials.

8. *What is the minimum distance at which the pins can be consistently perceived as two objects?*

   Answers will vary but should be in the range of 5 mm to 1 cm.

9. Since the subject expects to feel two pins, a possible source of error may arise. The subject's response may be biased. *How could using the rectangle with the single pin reduce this source of error?*

   Answers should suggest that a single pin should be occasionally used so that the subject does not know what to expect.

10. Using all six touch-testers, repeat steps 7 through 9 for each area of the body indicated in the table.
11. Exchange roles and repeat the tests.
12. Use your data to make a bar graph showing the average distance between touch receptors.

Analysis

1. **Summarizing Data** Summarize your data.

   Answers will vary but should describe a variation in distances between touch receptors for different areas of the body.

2. **Analyzing Data** Account for differences between the data obtained for you and for your partner.

   There is likely to be some variation in data between subjects, but the pattern of distances between body areas is likely to be maintained.

3. **Applying Methods** Describe a situation where expectations could produce erroneous data.

   Answers will vary but should describe a situation when the expectations of the subject lead to an erroneous response. For example, a patient might think that a medication relieved the symptoms of an illness, even though only a placebo was administered.

Name _____

Date _____ Class _____

A29 Graphing Growth Rate Data

Students compare data for height versus age in human males and females by making a graph. An analysis of the data reveals a similar pattern of growth for both sexes before and after puberty begins, which suggests that sex hormones may play a role in the process.

Background

Have you ever glanced at the family photo album and considered the developmental changes your body has undergone? If so, you have probably wondered what biological event triggered these transformations. Chemical messengers within your body dictated these changes. In this activity, you will analyze the relationship between age and human growth.

Objectives

In this activity you will
• **graph** and **analyze** human growth rate data.

Materials

• graph paper
• blue pencil
• red pencil

Preparation

Review the procedure for making a line graph.

1. Using graph paper, prepare a grid for graphing the data in the table shown below.

Age	Average height in centimeters	
	Males	Females
5	109	109
7	119	119
9	133	133
11	138	145
13	157	160
15	169	162
17	177	162
19	177	163

Procedure

2. Use a blue pencil to make a line graph that compares height to age for human males.
3. On the same graph, use a red pencil to make a line graph that compares height to age for human females.

Analysis

1. **Analyzing Data** Use the data to explain whether humans grow at a continuous rate, or in spurts.

Suggest that growth does occur in spurts especially around the age of puberty.

2. **Analyzing Data** Explain how the growth pattern for males differs from that of females.

The data indicate that growth rates for males and females are similar until puberty. The growth rate of females increases prior to that of males, as does the onset of puberty.

3. **Making Inferences** What evidence suggests that sex hormones might play a role in growth patterns?

The increase in growth rate at the time of puberty for each gender is evidence that sex hormones may play a role in growth.

4. **Evaluating Methods** How does the graph make interpreting the data easier?

The relationship between age and growth rate for both males and females is easier to observe in the graph than in the table of data.

HOLT
BIOSOURCES

LAB PROGRAM

QUICK LAB

A30 Collecting Data Through a Survey

This activity fosters group cooperation by having students conduct a survey relating to health and the use of tobacco, alcohol, and other drugs. This emphasizes student interaction and cooperation rather than concepts specifically relating to health.

Background

The advancement of science depends on the interaction and cooperation of human beings. This activity requires you to work together in a cooperative effort to collect data through a survey of your peers.

Objectives

In this activity you will
• work together to **design** a survey and **analyze** data.

Materials

• paper
• pencil

Preparation

1. Make a list of six statements that might be used in a survey to test a person's knowledge about health and the use of tobacco, alcohol, and other drugs. Three statements should be true and three statements should be false.
2. Form cooperative teams of four. Complete Steps 3–8.

Procedure

Students may write the statements during class or as a homework assignment. Encourage the students to consider relevant statements that truly test a person's understanding of important aspects of the topic. It is important that students research the statements. Emphasize the importance of writing well-researched statements to ensure that the survey is valid.

3. List your team's six statements on a sheet of paper.
4. Use your text or library references to check that half of the statements are true, and that half of the statements are false.
5. Your team should choose 12 statements for use in your survey. Have each member of the team make a survey data sheet similar to the model below.

Group: _____			
Statement	**True**	**False**	**Total**
1.			
2.			

6. Choose two different groups of people to interview. For example, you might compare students of different age groups, smokers and non-smokers, or adults and adolescents.
7. Have each member of the team interview three people from each group. Have each interviewee identify whether they believe each statement in the survey is true or false. Record each response in the appropriate column by making a check (✓).
8. Pool your team's data. For group 1, record the total number of responses and the number of correct responses to each statement. Then calculate the percentage of correct responses for each statement.
9. Repeat step 8 for group 2.

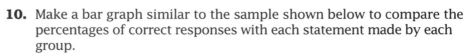

10. Make a bar graph similar to the sample shown below to compare the percentages of correct responses with each statement made by each group.

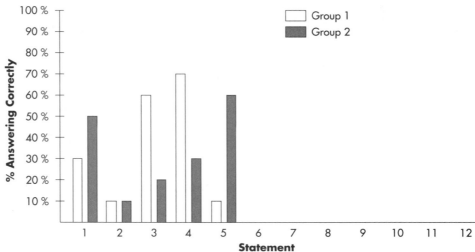

Analysis

1. **Summarizing Data** Summarize the data collected through your survey.

   Answers should adequately summarize the data collected through the survey.

2. **Evaluating Methods** List and describe the benefits of working in a group to complete this activity.

   Answers will vary but should discuss benefits such as the advantages of sharing in ideas and labor.

3. **Evaluating Methods** List and describe any problems that must be overcome when working in a group.

   Answers should discuss such problems as the difficulty in reaching a consensus and dealing with diverse personalities.

4. **Analyzing Data** Did you find any difference between the overall knowledge of the two groups?

   Answers will vary according to the data collected.

5. **Making Inferences** If any difference exists between the two groups, how would you account for it?

   Answers will vary according to the data collected.

6. **Analyzing Data** On which statement(s) did the subjects show the greatest need for information?

   Answers will vary according to the data collected.

Name _____

Date _____ Class _____

A31 Determining Lung Capacity

In this activity the students use a balloon to measure vital capacity, expiratory reserve, and tidal volume. Volume is calculated based on the diameter of the balloon (a sphere). While some error is expected using this method, the results will give a good estimation. Have students participate in this activity on a voluntary basis. Individuals with respiratory problems should not be forced to participate.

Background

Health workers accurately measure lung capacity using an instrument called a spirometer. These measurements provide one source of information about the general health of the lungs. In this activity, you will measure your lung capacity using a balloon. While this method is less accurate than a spirometer, it does provide a good indication of your lung capacity.

Objectives

In this activity you will
• **measure** lung capacity.

Materials

• round balloon
• metric ruler
• graph paper
• pencil

Preparation

Only round balloons should be used. For health reasons, have a fresh balloon for each student.

Review the technique for measuring the balloon's diameter. Remind students that the measurement of tidal volume should involve normal breathing only. Discuss the calculations involved in this activity. Some students may need help in making these calculations.

1. Form cooperative groups of four students. Work individually to complete steps 2–11.
2. Make a table similar to the one shown below for recording your data.

Trial	Balloon diameter measurements		
	Vital capacity	Expiratory reserve	Tidal volume
1			
2			
3			
4			
Total (all trials)			
Average diameter			
Volume cm³			

Procedure

3. Stretch the balloon to make it easier to fill.
4. Take a number of deep breaths. Exhale as much air as possible into the balloon. Hold the end of the balloon shut to stop the air from escaping.
5. Lay the balloon on the table next to the ruler. Measure its diameter in centimeters and record your data as *Vital capacity* for Trial 1. Release the air from the balloon.
6. Repeat steps 4 and 5 to complete Trials 2 through 4.
7. Breathe normally a number of times. Exhale normally. Then before inhaling again, force air in your lungs into the balloon. Measure and record the balloon's diameter as your *Expiratory reserve*. Repeat three times.
8. Breathe normally a number of times. Take a normal breath and exhale a normal amount of air into the balloon. Measure the balloon's diameter as your *Tidal volume*. Repeat three times.

9. Calculate the average diameter of the balloon for each group of measurements. Record the results.
10. Use the following formula to calculate the volume of air in the balloon.

Volume = $1.33 \, \pi \, r^3$

where r = $\frac{1}{2}$ of the average diameter of the balloon

and π = 3.14

11. Share your data with the other members of your group. Use all the data to make a bar graph comparing the three average lung volume measurements for each person in your group.

Analysis

1. **Summarizing Data** Summarize your data.

Answers will vary according to the individual.

2. **Defining Terms Operationally** Use your work to define vital capacity, expiratory reserve, and tidal volume.

Vital capacity is the maximum amount of air that is held in the lungs. The expiratory reserve is the amount of air that remains in the lungs after a normal exhalation. The tidal volume is the amount of air exchanged during a normal breath.

3. **Evaluating Methods** Describe any sources of error related to your final measurements of lung volume.

Answers will vary but could include sources of error relating to the measurement of the balloon, error associated with overcoming the resistance of the balloon, and errors in the calculations.

4. **Analyzing Data** How do your measurements compare with those of the other members of your group?

Answers will vary.

Explain the reasons for any differences.

Answers will vary but should indicate that body size, gender, age, physical condition, and the use of tobacco can contribute to variability between individuals.

An approximate measure of "normal" vital capacity based on height, age, and gender can be found using either of the following formulas.

Vital Capacity$_{MALE}$ in cm^3 = 52H − 22A − 3600
Vital Capacity$_{FEMAL}$ in cm^3 = 41H − 18A − 2690
where:
H = height in centimeters (without shoes)
A = age in years

Differences of 10 percent or less are not considered significant.
Have students compare the calculated vital capacity with their measured vital capacity.

Name _____

Date _____ Class _____

A32 | Relating Cell Structure to Function

Blood is composed of cells, cell fragments, and fluid plasma. Students use the microscope to examine some of the components of human blood. Based on these observations, they relate blood cell structure to function.

Background

Blood is composed of cells, cell fragments, and fluid plasma. In this activity, you will relate the structure of specific blood cells to the function each performs.

Objectives

In this activity you will
- **observe** blood cells.
- **relate** cell structure to function.

Prepared slides of human blood can be obtained from WARD'S. Alternatively, a 35-mm slide of a prepared blood smear can be projected for the entire class to use. Do not allow students to prepare slides of their own blood.

Materials

- prepared slide of human blood
- compound light microscope

Preparation

1. Form cooperative teams of two students to complete this activity. Use a separate sheet of paper to make a table for recording your observations.

Procedure

Review the basic structure of a cell. Draw the students' attention to the functions of the nucleus, cytoplasm, and cell membrane. Prepared slides may be obtained from WARD'S (93 M 6541).

2. Examine a prepared slide of human blood under low power of the microscope. Change the position of the slide and continue your examination.
3. Switch to high power and focus clearly on the cells. The most numerous cells are called red blood cells. Make a detailed drawing that shows a number of red cells. Label the parts of the cell that you recognize. Record your observations of the size, shape, and the relative number of the red cells.
4. Slowly move the slide until you find one or more larger cells that appear very different from the red cells. These cells are called white blood cells. Make a drawing of the white blood cell. Label the parts of the cell that you recognize. Record your observations of the size, shape, and relative number of white blood cells.
5. Look for evidence of tiny, dotlike cell fragments called platelets. Make a drawing of a few platelets. Record your observations of the size, shape, and relative number of platelets.

Analysis

1. **Identifying Relationships** The transport of oxygen throughout the body is one important function of blood. What evidence suggests that this function would be better performed by red blood cells than by white blood cells?

   The enormous number of red blood cells, compared with white blood cells, suggests that
   they function in the transport of oxygen throughout the body. The absence of a nucleus
   reduces the mass of the cell, while the concave shape increases the surface area.

_____

_____

2. **Making Inferences** What evidence suggests that only white blood cells are capable of reproducing?

The absence of a nucleus in red blood cells suggests that they are incapable of reproduction.

3. **Identifying Relationships** The capture and destruction of foreign organisms, such as bacteria, is another function of blood. What evidence suggests that white blood cells are better suited to this function than red blood cells?

The large size of white cells suggests that they are better suited for engulfing foreign particles. Also, the presence of a nucleus suggests that white cells have the ability to reproduce, which would be important in fighting invading organisms.

HOLT
BIOSOURCES

A33 Reading Labels: Nutritional Information

In this activity, students analyze nutritional information collected from food labels. Data for vitamin, mineral, carbohydrate, fat, protein, and caloric content are collected and organized in a table. This activity can be scheduled to introduce the unit or can be done after nutrition has been discussed.

Background

By law, the labels on food packages must list important nutritional information. Understanding and using this information can improve your health and can lead to better values in your food purchases.

Objectives

In this activity you will
- *analyze* the information listed on the labels of food products.

Materials

- 6 labels from assorted food products

Preparation

About one week before this activity is to be done, have students collect labels from a variety of food products. The amount of nutritional information recorded on a label varies from product to product. Extra labels should also be on hand in the classroom.

1. Collect labels from six food products. Be sure to include the portion of the label that lists nutritional information.
2. Make a table similar to the one shown below.

Food product	Vitamins	Minerals	Carbo-hydrates	Proteins	Fats	Calories

3. Form cooperative groups of four students. Work with one member of your group to complete steps 4 through 8.
4. Place the name of a food product in the first column of your table.
5. Use the "Nutrition Information" on the label to find the vitamin content of the food. List each vitamin present in the food in your table. Record the percentage of the recommended daily allowance (RDA) for each vitamin that you listed.
6. Record the mineral contents of the foods in your table.
7. Record the carbohydrate, protein, and fat contents.
8. Record the calorie content.
9. Repeat steps 4–8 for each of the food product labels collected by your group.

Procedure

Review with the class the nutritional information found on food labels.

Analysis

1. *Analyzing Data* In which foods do you find the greatest variety of vitamins?

   Answers will vary according to the data collected. _____

   _____

2. ***Analyzing Data*** In which foods do you find the greatest variety of minerals?

Answers will vary according to the data collected.

_____

3. ***Making Inferences*** Which foods are the most nutritious?

Answers will vary according to the data collected but should indicate foods that are higher in vitamins, minerals, and protein while being lower in calories and fat.

_____

HRW material copyrighted under notice appearing earlier in this work.

Name _____

Date _____ Class _____

HOLT
BIOSOURCES
LAB PROGRAM
QUICK LAB

A34 Culturing Frog Embryos

In this activity, students culture frog embryos and evaluate the effect of water temperature on embryo development. The exact temperatures used in this investigation are arbitrary, although temperatures below 4°C and above 28°C are likely to be lethal. To achieve a range of temperatures, embryo cultures may be placed in the refrigerator, in an incubator, and on a shelf in the classroom.

Background

Organisms in the early stages of development are called embryos. Scientists have learned that vertebrates share common genetic instructions for embryo development and are affected in a similar manner by changes within their environment. In this activity, you will culture frog embryos to study the effects of temperature on embryonic development.

Objectives

In this activity you will
• **evaluate** methods for culturing frog embryos.

Materials

• beakers (3), 100 mL
• pond water
• hot water
• frog embryos
• stereomicroscope

• watch glass
• culture dish
• wax pencil
• spoon
• thermometer °C

Preparation

1. Describe how water temperature might affect the development of frog embryos.

2. Make a chart similar to the one shown below for recording data.
3. **CAUTION: Put on a laboratory apron and disposable gloves.** Leave them on throughout this activity.

	Observations of tadpole development				
	Day				
Temperature	1	2	3	4	7
Room temperature					
Warm					
Cold					

4. **CAUTION: The hot plate produces high temperatures that can cause a serious burn.** Use the hot plate to warm the water. Use beakers to store warm, cold, and room-temperature water. Label each beaker accordingly.

Procedure

5. **CAUTION: Use humane treatment when handling live specimens.** Place frog embryos in a clean watch glass and cover them with pond water.

Observing the ongoing development of any embryo can be a formidable task. Discuss strategies for making a problem more manageable. The first step in managing a complex problem is to divide it into a number of smaller, simpler tasks. Observing and describing such specific features as size, shape, and apparent cell number should seem more manageable.

6. Use a stereomicroscope to observe the embryos. Notice the embryos' coloration. Record your observations. Make a drawing of your specimens.

7. Remove specimens that are severely injured or dead. *Why will removing these embryos benefit the remaining specimens?*

Answers should suggest that the dead or dying specimens are likely to increase the bacterial and fungal contamination of the water.

8. Label a culture dish "Room Temperature." Add room-temperature pond water to a depth of about 2.5 cm. Label a second culture dish "Cold" and a third "Warm." Add cold water to the second dish and warm water to the third dish. Record the temperature of the water in each of the three dishes.

9. Use a spoon to gently transfer 10 embryos to each dish.

10. Store the dishes according to your teacher's instructions.

11. Use the stereomicroscope to observe your specimens on days 1, 2, 3, 4, and 7. Record any changes in the embryos and make drawings. Remove any injured or dead specimens. Replace the water in each dish with fresh water, always at the same temperature.

12. Clean up your materials and wash your hands before leaving the lab.

Analysis

1. ***Summarizing Observations*** Summarize your observations.

Summaries will vary according to the data collected.

2. ***Evaluating Methods*** What temperature is best for culturing frog embryos?

Answers will vary according to the development of the cultures but might state that embryos in the cold culture appear to develop more slowly than those in the room-temperature culture.

Explain why.

The warm culture may develop more quickly but with more fatalities.

3. ***Making Inferences*** How might the jellylike layer that surrounds the embryo help it survive in water that varies in temperature?

The jellylike layer provides insulation, reducing fluctuations in the embryo's temperature.